"The book is proof that the ridiculous and the horrible are joined at the hip and will never be separated; in a universe that seems to continually be thinking up new ways to kill us, Brockway knows there is but one rational response: to point at it and laugh."

—David Wong, author of *John Dies at the End*

"There has never been a hipper Prophet of Doom. Irreverently entertaining and terrifyingly accurate, the probability of our future death has never been more fun."

—Michael Largo, author of *Final Exits: The Illustrated Encyclopedia of How We Die*

"This book is hilarious. As if I wasn't already afraid to leave the house. What's that? Brockway suggests the apocalypse doesn't knock? What if I cower under a desk? Well, damn. I might as well have another nacho."

—Garth Sundem, author of *The Geeks' Guide to World Domination*

EVERYTHING
IS GOING TO KILL
EVERYBODY

ROBERT BROCKWAY

EVERYTHING IS GOING TO KILL EVERYBODY

THE TERRIFYINGLY REAL WAYS THE WORLD WANTS YOU DEAD

THREE RIVERS PRESS NEW YORK

Published in the United States by Three Rivers Press, an imprint of the Crown Publishing Group, a division of Random House, Inc., New York.
www.crownpublishing.com

Three Rivers Press and the Tugboat design are registered trademarks of Random House, Inc.

Library of Congress Cataloging-in-Publication Data
Brockway, Robert, 1980–
Everything is going to kill everybody : the terrifyingly real ways the world wants you dead / Robert Brockway.
p. cm.
Includes bibliographical references.
1. Hazardous substances. 2. Natural disasters.
3. System failures (Engineering). I. Title.
T55.3.H3B76 2010
303.48'5—dc22
20090393312

ISBN 978-0-307-46434-7

Printed in the United States of America

Design by Maria Elias
Illustrations by Jason Pedegana

1 3 5 7 9 10 8 6 4 2

First Edition

This book is dedicated to Meagan, who has had the kindness and decency to not realize she's way out of my league for many years now. It should also be noted that this entire book was her idea. Please direct complaints accordingly.

CONTENTS

BIOTECH THREATS

ROBOT THREATS

INTRO

There is no shortage of modern apocalyptic thought: Some Christians thought that the second coming of Jesus would coincide with the turn of the millennium, for technophobes the Y2K scare was the secular equivalent of Armageddon, and most cults declare the end-times every other Wednesday (and alternating Saturdays), while the new hot apocalypse theory on the block, based on ancient Mayan prophecies, says that the end of days is set for 2012.

These are just the current dates, but humanity has believed that we are living in the end-times since the *beginning* of time. From the moment humans gained sentience, we immediately thought that everything around us was irrevocably fucked, and that it was just a matter of years—if not days—before the whole world came crashing down around us. Nostradamus is the most famous of these ancient criers of doom, but he was not the first by far: One of the earliest apocalyptic heralds is recorded in an Assyrian clay tablet dating all the way back to 2800 BC, which reads, in part:

> Our earth is degenerate in these latter days. Corruption is rampant. These are signs that the world is speedily coming to an end.

This tablet, in addition to being one of the first recorded apocalyptic prophecies, also holds the dubious honor of being the oldest inaccurate prophecy in history—considering that since you're alive to read these words, that anonymous author was wrong by at least five thousand years. Though there have certainly been other mistaken prophets of doom since, none has been such a truly epic failure. It just goes to reinforce that age-old adage: You can't spell Assyrian without an "ass."

The end of days isn't just reserved for cultists and tablet-dictating prehistorical idiots, however. The presumed apocalypse has held intense fascination for people from all walks of life. Sir Isaac Newton, for example, was obsessed with eschatology, the study of Armageddon, and he spent a good chunk of his life attempting to "decode" the Holy Bible, writing a five-thousand-word treatise on the subject. He was absolutely certain that the Bible contained an underlying cryptic code that, if unlocked, would reveal all the secret laws of the universe. After apparently finding his key in the Book of Daniel, Newton settled on the year 2060 as the official RSVP date for the end days.

This is a little-known aspect of the famous pioneer of science, because he was loath to show his work to anybody; this obsession was only recently uncovered in private manuscripts by scholars at the Jewish National University Library Archives in Jerusalem. Newton was so sure of his findings that he even started a preindustrial beef with other apocalyptic prophets, stating that:

> It may end later, but I see no reason for its ending sooner. This I mention not to assert when the time of the end shall be, but to put a stop to the rash conjectures of fanciful men who are frequently predicting the time of the end, and by doing so bring the sacred prophecies into discredit as often as their predictions fail.

That's some serious trash talk, Newtron Bomb! You not only insisted that everybody but you is wrong, but that they are all "fanciful men" (which is how people questioned your sexuality before the internet made it so easy just to type "fag") who fail so hard that they disgrace God himself. But before you go thinking that one of the most notorious geniuses in history may have actually cracked the mathematical code of the Almighty, you should also know that Newton believed the apocalypse would be preceded by a thousand-year reign of immortal saints—of which he was one. So, unless you believe every day of the last thousand years has been heaven on Earth and that Isaac Newton is basically the Highlander—secretly ruling the world from his immortal apple orchard—perhaps you shouldn't put too much stock in this particular hypothesis. There is, after all, a fine line between genius and insanity—and apparently Newton wasn't a "color inside the lines" kind of guy.

Don't start believing that foretellers of the apocalypse are somehow relegated to the annals of ancient history, either. It wasn't all Nostradamus, Newton, and ancient Mayans: In 1910 there was an *extremely* popular theory that the passing of Halley's Comet would somehow poison the entire Earth with cyanide gas. This was so common, in fact, that snake-oil salesmen of the day made fortunes from hawking so-called Comet Pills that would ostensibly render people immune to unearthly toxins.

Sure, mostly it's just whackjob cultists who are preaching the end-times these days, but it's easy to see the appeal of the concept: As human beings we are all keenly aware of our own mortality, but although we know we all have to die eventually, there is some small amount of comfort in knowing that maybe it's something we could all do together, as a team. There is, after all, no "I" in "apocalypse."

And so, in the same spirit of these joyously stupid celebrations of the potential end of humanity, this book is going to take you through all the very real ways that we, as a species, have already damn near died, may be dying off right now, or may be suddenly wiped out in the future. What this book will not do is take any sort of metaphysical stance on the end of mankind, by detailing cop-out bullshit that could "irrevocably alter our society." This book will not take a "grand scheme of things" approach by covering stuff like astronomical events billions of years in the future that could destroy our long-dead planet. And this book will absolutely not take the easy route, by simply fictionalizing some potential apocalyptic scenarios and asking "Wouldn't that be scary, folks?" No, every single subject within this book fits three simple, important criteria: They affect the reader directly, they pose a real threat, and they could happen soon. To put it more succinctly: Everything in here will kill you and everyone you love in various horrible ways, and there's not much you can do about it but laugh. Or void your bowels and call your mom. It's your choice, really (but the former option is substantially less embarrassing and slightly less disgusting).

NEAR MISSES

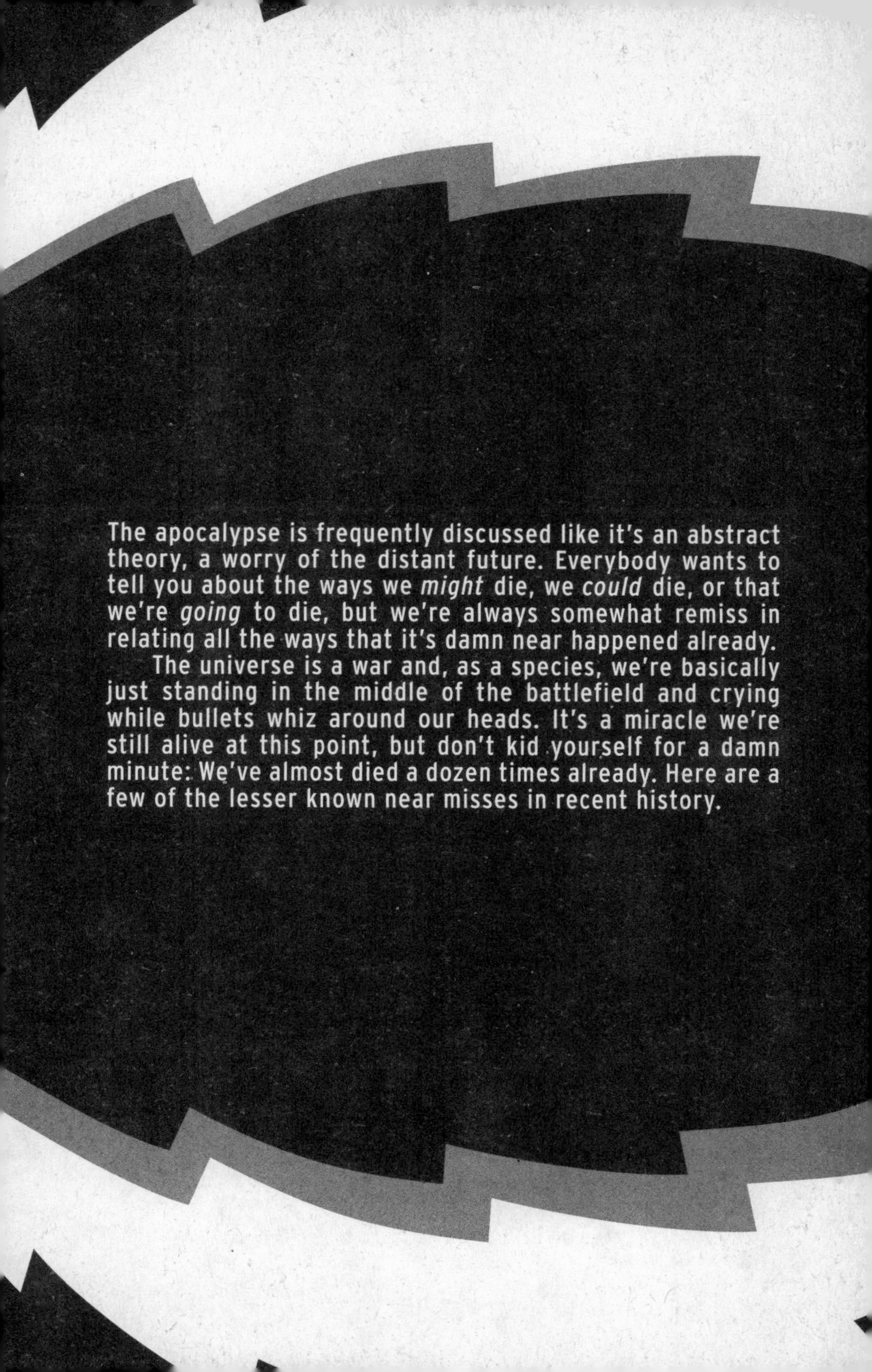

The apocalypse is frequently discussed like it's an abstract theory, a worry of the distant future. Everybody wants to tell you about the ways we *might* die, we *could* die, or that we're *going* to die, but we're always somewhat remiss in relating all the ways that it's damn near happened already.

The universe is a war and, as a species, we're basically just standing in the middle of the battlefield and crying while bullets whiz around our heads. It's a miracle we're still alive at this point, but don't kid yourself for a damn minute: We've almost died a dozen times already. Here are a few of the lesser known near misses in recent history.

1.

STANISLAV PETROV

IT WAS JUST PAST midnight in Russia on September 26, 1983, and the Cold War was at its coldest . . . and warriest. A recent transgression by Soviet military forces had left U.S.-Soviet relations more tense than a Sammy Hagar/ David Lee Roth threesome—that is to say, somebody was getting a dick in the eye; it was just a matter of time.

The "transgression," in this case, consisted of Soviet fighter jets blowing a Korean Air Lines passenger plane straight out of the sky. Two hundred sixty-nine people died in this incident, including one Larry McDonald, United States Congressman. Considering that we're talking about the height of the Cold War here—where a windblown fart would have been reason enough to nuke a continent—that's a pretty big "incident." The fighter pilot's justification for exploding a small town's worth of people flying in the danger equivalent of a giant retarded duck? The plane maintained radio silence when hailed. Some might call that a "holy fuck-ton of overreaction" just for getting the cold shoulder from a commercial airliner, but you must keep in mind that Russia at the time was a highly volatile place. These kinds of overreactions were probably common in the USSR, leaving hot-blooded young Russian males so high-strung that, upon receiving the cold shoulder from anything—even girlfriends—a reasonable knee-jerk response was to immediately fire high-yield explosives at the offending woman until she plummeted from the sky in flames . . . probably.

On top of all this preexisting tension, a NATO exercise was under way in Europe. Operation Able Archer had temporarily raised NATO nuclear alert levels in preparation of a simulated nuclear war. To help put the sheer, palpable levels of death in the air in more relatable terms, let's use us a down-home analogy: Let's say you and your neighbor don't get along. Never have, never will. Such is life. But one fine day, your wife wanders onto the neighbor's sidewalk, whereupon she is immediately hit with a hand grenade by said neighbor, who then runs up and down the border of your two lawns screaming obscenities and insisting that you "don't have the balls" to do anything about it. Also, you happen to have been outside this entire time, conveniently polishing your collection of machetes and dynamite. Pretty much everybody is going to die here—it's practically destiny.

And so was the mood when Lt. Col. Stanislav Petrov took charge of Serpukhov-15, a Soviet satellite station monitoring the skies for signs of nuclear attack. Petrov had agreed to cover the night shift for a coworker, so he was there when a blip suddenly appeared and the klaxons sounded. The long-distance radar was picking up a nuclear missile launch from America, targeting Moscow. While his fellow officers most likely shit themselves with fear and began slapping at nuclear launch buttons like a bunch of epileptic seamstresses, Petrov calmly deduced that the missile was likely a computer glitch.

Top Four Worst Things to Sound

- Puget
- Of Silence
- Of Music
- Klaxons

And luckily the right man was in charge. Petrov literally wrote the book on pants-wettingly terrifying near apocalypses; he was actually the author of the manual dictating Russian military response in the event of a perceived nuclear attack. So, being a key member in the development of the early-warning system, he happened to know firsthand that it was a relatively new, untested, and inherently flawed system, little better than a bunch of bells duct-taped together and shot into space to ring if something hit them. Still, it was the only system they had, and it was Petrov's duty to warn the government of whatever missile attacks it registered, so that they could retaliate before they inevitably died. Because that was all one could do in the event of a full-blown nuclear attack: Not protect yourself, not surrender, not evacuate—just make sure that whoever fucks you gets a good solid boning in return. That's called Mutually Assured Destruction, and it was the only thing left to do except cry in Petrov's situation.

But he held fast. "Why would the Americans launch only one missile," Petrov reasoned, "doing comparatively little damage, but provoking a full-scale counterattack from the Russians?"

If the perceived attack was a glitch and they launched missiles in response, the Americans would launch *their* response for real, and the resulting nuclear holocaust would kill millions on both sides. Petrov called a halt to the Mutually Assured Destruction and waited. Seated next to his enormous, steel-clad balls, he stuck by his instincts and waited—waited to see if everyone and everything he loved was about to be turned to ash by virtue of his hunch.

MAD stands for . . .

A: Mothers Against Dockworkers
B: Muppets Are Demons
C: Malicious and Delicious
D: Mutually Assured Destruction

And then it got worse! Short of getting sudden, massive, incurable rectal cancer, how the hell could this situation get any worse, you ask? The man is already waiting for a nuclear missile to strike, gambling his human instincts against the cold unerring judgment of a computer with the life of his entire country on the line—how much worse could it possibly get?!

Five times, actually.

That's an oddly precise number, isn't it? Five times worse; one for each missile that seemed to be launched. One after another, five blips followed suit and appeared on his screen. One missile blip's a fluke, OK, but surely *five* Minuteman intercontinental ballistic missiles showing up on screen, one after another, is a pretty solid omen of the apocalypse. Why not hit the "well, fuck you too, buddy" button and at least go down swinging?

Because Petrov is a stone-cold motherfucker, that's why. In his own (absurdly calm, all things considered) words, "Suddenly the screen in front of me turned bright red. An alarm went off. It was piercing, loud enough to raise a dead man from his grave. For fifteen seconds, we were in a state of shock. We needed to understand, what's next?"

That's right. Read that again: fifteen seconds.

I spend ten times longer than that deciding what socks to put on in the morning, and that's how long it took Petrov to a) process the very real potential of an apocalypse, b) psychoanalyze just how hard-core the Americans' death wish was, and c) reason out a never-before-seen computer glitch in the Soviet Nuclear Defense system, whereupon he decided to place his entire career on the line in order to save the Earth. In fifteen seconds, Petrov made his assessment, and in three minutes, he would find out if he was right.

You just spent as much time reading the recap of his actions as Petrov did saving the world. But though the decision was made quickly, the wait was impossibly long. As the word START ominously flashed bright red on his terminal, practically begging him to reconsider, Petrov stuck by his decision not to retaliate for three eternal minutes. Three minutes of waiting to find out whether the nuclear missile heading for his face was real. Even if they were waiting in complete and total silence,

More Things You Can Do in Fifteen Seconds

- Tie your shoes
- Brush your teeth (if you're gross)
- Disappoint a woman

three minutes is a tense and terrifying length of time to consider your own death. But they weren't waiting in tomblike quiet: If the system registered more than a single rocket, it was programmed to send immediate word to headquarters. Which kind of makes you wonder what the purpose of Petrov's whole Atlas-like "only you can decide if the apocalypse is real" duty was, if the computer just goes straight over his head the minute shit gets real anyway, like a little brother in a headlock calling for Mommy. So instead of respectful silence, Petrov spent those three minutes arguing with a panicked man over the phone, screaming commands into the intercom, and, knowing Petrov, probably winning a quick game of Heads-up Basketball while banging eight supermodels in a Ferrari.

System checks kept being run on the computers, and everything kept coming back not only positive, but also registering the highest level of accuracy in its assessments. "Everything's cool," the computer whispered, "just kill everybody and we're totally cool."

Because you're not a dramatic silhouette burned into a sidewalk right now, you already know what happened: nothing. The "launches" were false alarms, glitches in the system caused by stray sunbeams registering as missiles (although the fact that the Russians' elite military systems confused a sunny day for a nuclear holocaust and nearly wiped out the Earth in retaliation probably does little to set your mind at ease). The system was actually designed specifically to rule out this exact effect, but due to a near-cosmic alignment of the sun, U.S. missile fields, cloud coverage, and atmospheric phenomena, it was tricked into doing the one thing it was built to avoid, making any possibility of this particular error occurring practically nil. And still Petrov called it out.

If this were poker, Petrov would've just bluffed God out of his winning hand without even holding cards. When asked about his personal opinion regarding such an impossibly fucked situation, Petrov laconically replied that it was all just "God's own little joke of outer space."

I hope he put on sunglasses while electric guitar solos raged behind him after a line like that.

So regardless of the situation, he saved the world! He's a hero, right? Nope. Not exactly. As stated before, if those satellites registered an attack, Petrov's only duty was to hit that retaliation button and murder a continent. The entire human race, Russians included, is alive today only because this particular Soviet soldier made an impossible judgment call. What happened to the man who saved the world?

His superiors were so embarrassed that there was even a problem in the first place that they began a "formal investigation" into Petrov's "failure of duty." Among other things, they were quite perturbed that Petrov had not taken notes during this ordeal. They insisted that, between talking down Russian Nuclear Tactics headquarters from their kill frenzy, personally assessing the viability of the most advanced technology on Earth, and playing chicken with the angry hand of God, Petrov should have also shown his work, like a grade-school algebra student. Why didn't he, they asked?

His exact, word-for-word response:

> Because I had a phone in one hand and the intercom in the other, and I don't have a third hand.

When confronted with such infallible logic by the man solely responsible for the continuing lives of your children,

> **Stanislav Petrov's Shown Work**
>
> **Step 1:** Maintain calm in face of Armageddon.
> **Step 2:** Call computer on its bullshit.
> **Step 3:** Divide by GIANT BALLS.
> **Step 4:** Stand by hunch with fate of humanity hanging over head.
> **Step 5:** Carry the balls.
> **Step 6:** Save world.

what did the higher-ups do? They reassigned him to a "less sensitive" position, from which he soon retired. He received a pension of about two hundred dollars a month, and the record of this incident was kept top secret until 1998. That's fifteen years of complete obscurity. Fifteen years of knowing you saved the world, and getting the career equivalent of being reamed in a broom closet for your troubles, and somehow Petrov still soldiered on. During this period of obscurity, he suffered a series of debilitating tragedies: His wife died, his legs became so badly infected that he couldn't walk for months at a time, and the village he lives in wouldn't even give him a job sweeping the streets.

So, to recap: He was basically fired, for not *completely* fucking over his own government. Because in Soviet Russia, government fucks you.

Since 1998, however, when word of his astounding deeds was finally made public, he has received a World Citizen Award! Amazing! It may have taken forever and a day, but a good deed is rewarded!

With one thousand dollars and a trophy!

That's not a typo. One thousand dollars. And a trophy. The same reward for winning a regional amateur bowling tournament, for the man who saved the world. Thanks for not ending humanity; buy yourself an '83 Honda Civic.

But what does Petrov himself have to say about his pivotal role in the continuing existence of human life as we know it? What does he think about his own actions, the fate of the Earth, and the subsequent disbarment and tragically inadequate rewards?

"I was simply doing my job and I did it well. Foreigners tend to exaggerate my heroism."

Well, shit. Let's hope he never reads this. He'd probably kick my ass.

2.

KLEBSIELLA PLANTICOLA

IN THE 1990S, A European biotech company prepared to commercially release a genetically engineered soil bacterium for use by farmers. They were operating under two very reasonable assumptions:

1. Nobody likes plant waste.
2. Everybody likes booze.

Whereas the common man might address these issues by simply not doing any plowing and opting to *get* plowed instead, scientists at the biotech company thought of a much more elegant solution: Engineer a bacterium that aggressively decomposes dead plant material—specifically wheat—into alcohol. And in 1990, they did exactly that. The bacterium was called *Klebsiella planticola,* and it nearly murdered everybody; you just don't know it yet.

Klebsiella planticola is of the enterobacterium family, microbes that typically reside inside the guts of mammals, but this particular strain inhabits the root systems of most terrestrial plants. Actually, every root system that's ever been tested for the presence of *K. planticola* has come up positive, so it is as near to a universal plant bacterium as there has ever been (you should remember that part, because it's going to come in handy later). In its premodified, natural form, *K. planticola* is partly responsible for the decomposition of all plant matter—a vital step in the natural life cycle—and it's notoriously aggressive in this role. That's why it was picked out for experimentation in the first place: Like an Old Testament God, *K. planticola* is both omnipresent and incredibly belligerent.

Biotech researchers saw these traits and thought they seemed perfect for an agricultural problem they were working on. Burning off dead plant material, as was the standard practice, severely pollutes the air and damages the lungs of farmers. But what if, instead of the regular old largely useless sludge that decomposing plant material results in, we could alter that sludge into something more useful to humans, thus eliminating the desire to simply burn it away? What if we could ferment it, and turn it into an alcohol, a fuel, or a hyperefficient fertilizer? Or better yet, all three! Why not get blitzed off of it, piss it into your gas tank to power your car, and then puke it up into the yard to make your garden grow?

Suddenly alcoholics are useful members of society again. Hell, they're practically heroes: brave men and women sacrificing both their livers and their dignity to bring us power, food, and alcoholic-inspired confidence! Well, that's the noble goal biotech researchers had in mind when they spliced an alcohol-producing bacterium into *K. planticola.* Once their product was released, farmers would simply gather the dead plant matter into buckets and let it ferment into alcohol. Alcohol that could do everything they hoped: Be distilled into gasoline, sowed as fertilizer, burned as cooking fuel, or just drunk by the filthy, dirt-tasting bucketful. Their bioengineered *K. planticola* would create a beautiful, Eden-like garden paradise.

So it was all with the intent of doing good that they engineered this microbe, but you know what they say about "the best intentions," don't you? That's right: They inevitably result in pestilent, humanity-destroying plagues.

See, it was that fertilizer part where things got, shall we say, fucking horrifying: Once the fermentation process necessary to turn that dead plant material into alcohol occurred, the sludge left over would be rich in nitrogen and other such beneficial substances, making it an ideal fertilizer. The plan was to spread this sludge fertilizer back on the fields, thus eliminating all waste from the whole process.

Biotech Scientists

"What if we could use plant waste as fertilizer?"

"Yeah, and what if we power our cars on it?"

"Hey, yeah! And what if we could get fucked up on it too?!"

". . ."

"I think you might have a problem, Ted."

An Exchange from Utopia

"Honey, did you mow the lawn yet?"

"Fuck yes, baby. Where you think I got this weed whiskey?"

"I love our perfect lawn, darling!"

"I love free booze!"

The big problem? The fermentation process didn't kill the modified *K. planticola*—it was still there, ready to turn dead plant material into alcohol.

The bigger problem? It didn't even wait until the plants were dead to start.

The normal *K. planticola* bacterium results in a benign layer of slime on the living root systems it inhabits, but the engineered version would also be producing alcohol in this slime—with levels as high as seventeen parts per million, and anything beyond one or two parts of alcohol per million is lethal to all known plant life. So the engineered *K. planticola* basically gives all plant life it touches severe alcohol poisoning, putting them more than ten times over the lethal limit of fucked up. Like a frat house during pledge week, *K. planticola* would force all new plants it encountered to drink well beyond their reasonable limits. But unlike frat-house rushes, it's not just freshman idiots who are affected, it's everybody. So maybe that analogy isn't entirely accurate: It's more like a bleak dystopian future where frat houses rule the world with a tyrannical fist, hazing and beer-bonging humanity into the grave. Because, you'll remember, *K. planticola* is present in all plant life.

Every species.
Every variety.
Poisoned.
To death.

Now those wonderful traits that made it such a good candidate for modification in the first place—its notorious aggressiveness and near omnipresence—are no longer such good things, are they? Because if there's one thing you really don't want your poison to be, it's "notoriously aggressive." And if there's one place you absolutely do not want your "notoriously aggressive poison" to be, it's *everywhere*.

Keep in mind that this was not a theoretical scenario, far-flung, fictional, and unlikely to ever actually occur. This bacterium was going to be released; it had all of the necessary approval. It was only a matter of proper marketing and shipping at this point. It was only by virtue of a random review by an independent scientist (Dr. Elaine Ingham, a professor at Oregon State University and possibly the savior of all mankind) that it was caught in time.

How the fuck could this possibly happen? How did the leading biotech researchers of the day not realize that they had engineered a bacterium that would kill all plant life it touched? Did they not test it on any, y'know, *plants?!*

Well, for all intents and purposes: No, they didn't.

See, the Environmental Protection Agency was the only overseer for all biotech releases, and their policy was to test new bacteria in sterile soil. The problem here being that the real world is not sterile; it is the antithesis of sterile. The whole point of sterility is to zap all normal, unexpected elements out of a sample environment so that the scientists can see its effects in a pure, untainted environment. They deemed the modified *K. planticola* to be safe in sterile soil,

Excerpt from EPA Protocol Manual

If your dirt is dirty, make sure to take all of the dirt out of the dirt before you test the dirt.

but apparently just totally forgot that its intended use was in the fucking dirt, which is a notoriously *dirty place*, isn't it?! Luckily, Ingham and her group took it upon themselves to study the bacteria in a more realistic scenario, using normalized samples of unsterile soil and three different sample groups. There was a group absent of *K. planticola* entirely, a group with the normal *K. planticola* present, and a group with the genetically modified *K. planticola* in it. They planted wheat seeds in all three groups, and then let it sit for a week. When they came back they found the first two groups doing fine, while all the crops from the GM sample were dead.

Dead in less than a week.

If released from the lab—which, I cannot stress enough, it very nearly was—the modified *K. planticola* would have spread worldwide in a matter of months, killing all plants it touched within a week, and turning all soil-based plant life into sweet, sweet liquor. Like a twisted hillbilly fantasy, the world really very nearly drowned in moonshine.

CURRENT THREATS

A lot of speculation about the apocalypse is based in the worries of an uncertain future: Will that meteor hit us? What new threats will technology bring? What crazy diseases will we fight in the future?

But there's no need to *wait* to get that sick fear-fix you so desperately desire!

There's plenty of shit going down *right now* that may already be wiping the human race off the face of the Earth as we speak! Don't dwell in the future or the past, friends! After all, the reason they call it "the present" is because every day is a gift! A terrible gift that you most certainly should not unwrap. (If you shake it, it sounds a lot like genocide.)

3.
FRANKENCROPS

GENETICALLY modified foods are all the rage, and recent advances in technology have us doing some pretty crazy stuff, like splicing antifreeze fish genes into tomatoes, hamster genes into tobacco, or even chicken genes into potatoes! And while fish tomatoes and hamster cigarettes sound a little disturbing, who can't see the appeal of a chicken potato? They put peanut butter and jelly into the same jar; why not apply the same thinking to KFC? If you ask me this is just a case of science doing stoners a solid, but some killjoys decided to start looking into potential side effects of these pimped-out foodstuffs, and to the surprise of nobody, they've found that we're already completely fucked to the gills . . . much like a fish tomato.

In general, genetically modified crops tend to be thought of as a slowly developing problem at worst. Considering that if you're willing to pay a bit more, you can currently find organic pretty much anything, natural food is still readily available for any and all takers—provided you hate disposable income like it killed your father. But just because organics are available to elite folks in industrialized nations doesn't mean that's true for the developing world. When you also consider that the entire worldwide agrochemical market is owned by just ten companies—the entire *thirty-billion-dollar-a-year market*—you start to see signs of a future monopoly. Two companies in particular seem ripe (zing!) to govern a dystopian future straight out of science fiction: Novartis and Monsanto. Novartis is the less worrisome of the two, and—considering that they're a single massive corporation that controls the creation of plants, the distribution of food crops across the world, and the medicines that keep you alive—being "less worrisome" is really saying something for the terror factor of their rival global-domination food company, Monsanto. See, Monsanto doesn't seem content with meager ambitions like ruling the world's food and medicinal supplies—they're aiming to control the very nature of food itself, with something they have not-so-comfortably dubbed "Terminator Technology."

Really.

That's not a nickname made up by their opposition; Monsanto themselves named their product "Terminator." It's like they're flaunting their potential supervillainy! In the realm of terror-inspiring corporate decisions, that's right up there with naming your headquarters Death Mountain, and awarding Employee of the Week to Hitler's-Brain-in-a-Robot. Terminator Technology isn't just an unfortunate name, however: It really is every bit as worrying as the moniker implies. It refers to genetically modified plants that produce only sterile, dead seeds, and so cannot ever reproduce naturally. Monsanto hopes to eventually replace all of its agricultural seed sales with this technology, thus forcing all farmers to purchase new seeds from Monsanto yearly, since they'll be unable to simply plant the seeds gathered from the previous year's crop. The benefits to farmers using Monsanto's new seeds must be enormous, right? Not so much! Lower prices and a slightly more stable crop. That's pretty much it.

Better Names for Sterility-Based Plant Products

- Red Barren
- Infertile Foliage
- Impotent Potables

Oh, but the perceived benefits to Monsanto? Nothing less than the complete ownership of plant life. That sounds like the plot point to a fucking *Captain Planet* episode; surely it could never come to pass! The government surely must be watching and making sure that no one single private corporation could own an entire fundamental necessity of life, right? That's like Microsoft buying "air," or Walmart owning the patent for "shelter." It's so absurd that there's no way it can slip by the authorities. And it hasn't.

The government knows all about it.

They've heard about it from Michael Taylor, for one, an attorney and proponent for Monsanto as well as a current employee. He told the government all about Monsanto's sinister plans . . . when President Obama appointed him to the Food Safety Working Group in 2009. Or maybe they first heard about Monsanto's bid for domination when Michael Taylor was Deputy Commissioner for Policy at the Food and Drug Administration in the early 1990s, or maybe when he was Administrator of the USDA's Food Safety and Inspection Service in the mid-'90s. This guy is currently one of the central points of control for the government's oversight of genetically modified foods, and he is working for the company that wants to own the very concept of food.

So yes, the feds are well aware of Monsanto's aspirations, and they think that's just awesome. They probably wish they'd thought of it first, if anything.

Terminator Technology is particularly worrisome because it *forces* farmers to pay for next year's seed every year. So if there's any fluctuation in a farmer's income—due to drought, infestation, or other unforeseen circumstances—and he can't pay, Terminator Technology basically functions as a guarantee that there will be no crops to make up for it the following year. It's like a ticket to go fuck yourself next season!

Act Now and Get 30% Off the Return Trip!

*Return trip does not reverse any pestilence, famine, or plague resulting from original ticket.

Valid every single year! Forever!

But hey, if you don't buy this shit because you, like a sane person, are not exactly enthusiastic about a corporation renting your own belly out to you, then surely it doesn't affect you. . . .

Wrong again! Man, it seems like every rhetorical question you ask this book turns out to be wrong. It affects everybody, according to experts like Camila Montecinos, a Chilean agronomist, who says:

> We've talked to a number of crop geneticists who have studied the [Terminator] patent. They're telling us that it's likely that pollen from crops carrying the Terminator trait will infect the fields of farmers who either reject or can't afford the technology. Their crop won't be affected that season but when farmers reach into their bins to sow seed the following season they could discover—too late—that some of their seed is sterile.

So if you, as a farmer, live near anyone who buys Terminator Technology, well, Terminator Technology then comes for you. And like Kyle Reese (who is no biologist but does have a doctorate in starring in the movie *Terminator*) says: "Listen, and understand. That terminator is out there. It can't be bargained with. It can't be reasoned with. It doesn't feel pity, or remorse, or fear. And it absolutely will not stop, ever, until you are dead."

Don't tell me Kyle Reese isn't an expert. The same principles from the movie apply to this situation: Just replace the robot with a modified gene, Linda Hamilton with all plant life on Earth, and your life with, I don't know, that biker Arnold murders in the beginning. You don't last long, is the general point.

OK, so even assuming all of this is going down like I said, it's not like corporations are inherently evil. We're all human, after all, and even the largest corporation has to have some sort of good in them. Well, if you're still willing to trust in the moral equilibrium of a company that names its products after death machines from the future, here's a fun fact for you: Monsanto started out manufacturing artificial sweeteners, but eventually started taking on government contracts, where they developed Agent Orange, the poorly tested, hastily deployed herbicide-turned-chemical-warfare agent used in the Vietnam War that poisoned and killed thousands of enemy combatants and civilians as well as sickening our own troops.

Who better to govern all food on the planet than a corporation previously versed in plant-based warfare? The only people who don't realize that's a sarcastic question are apparently running the U.S. government.

But what if all plant life on Earth doesn't want to just quietly die off? It may well take the other track: genetically modified foods that destroy the world through aggression and growth. In Alberta, Canada, there are canola, or rapeseed, plants that have become resistant to all three of the most commonly used herbicides. This has forced the farmers to use a chemical called 2,4D, an extremely powerful pesticide, just to keep everything even, basically starting an arms race against Mother Nature, and she's bringing her A-game. Could this be the first worrying step in the rise of a sentient plant army, immune to conventional weapons and fueled by a vicious hatred of mankind that borders on genocidal insanity? Some experts say yes.

They're experts in Crazy, but that still counts as an expert (to crazy people).

All joking aside, though: Plants are going to murder your family.

Wait, no, for real this time: This increased immunity to herbicides is actually our fault. The biotech engineers modified the rapeseed crops *on purpose* with immunity to all three of our major pesticides in the hope that it would protect the crops from the harm done by pesticide spraying. The only worrying part? Unless you're specifically trying to grow rapeseed, it's one of the most virulent weeds around, and now it's damn near impossible to get rid of it when you want to grow something else, like, say, food. Anything without a genetically modified advantage like the rapeseed has gets, well . . . raped.

Crazy Experts Consulted

- Ray "Alley Meat" Johnson
- BORGO: ALIEN SUPPLANT
- Gary Busey

That's perhaps the clearest example we have of the unintended consequences of gene manipulation, but there's something more disturbing to consider. The term "gene flow" refers to the transfer of modified genes to unmodified plants. If a wild weed is closely related to a modified food crop, the genes from the altered plant can naturally flow via pollen and interbreeding into the unaltered weed, theoretically causing superresistant strains that, in turn, will naturally transfer the artificial genes through further cross-pollination to other relatives again. Eventually, introducing a pretty-much invincible plant could end up rendering everything remotely similar in genetic structure practically invulnerable as well. And this is not strictly a theoretical scenario. These "invincible" plants are being bred right now. There's a company named Ciba-Geigy that manufactures "Maximizer Corn," a crop that produces insecticide in every single cell. Every single cell is a tiny, deadly badass, burning through insects like action heroes burn through henchmen. What if that gets into crop-destroying weeds, or worse, harmful or even poisonous plants? I'm not saying you could end up losing a fistfight to a shrub, I'm just saying you may want to be working the bag a little, just in case.

But medicinal advancement works both ways: That simple boost in immunity can start screwing over plants exactly the same way that it does us. We start overusing antibiotics and as a consequence we start to see the evolution of superviruses that cannot be stopped. Similarly, we boost plant immunity, and new superbugs start ravaging crops and we have no defense against them. We're basically teaching plants how to use biowarfare against themselves . . . and it's about goddamn time! Why should corn live in peace when we must live in terror? Fuck you, corn. We'll genetically engineer you to feel fear if we have to.

We've genetically modified *millions* of plants that are currently crosspollinating totally at random and completely out of our control; who knows what monstrosities will emerge from that? When you think about it, it's really only fair for us to screw over the plant world. Even if we did start this disturbing battle, unleashing biblical plagues on plant life is practically self-defense now. It's not a matter of *if*, but *when* the chicken potatoes come for you, and may God have mercy on your soul should they find you with your fryer cool.

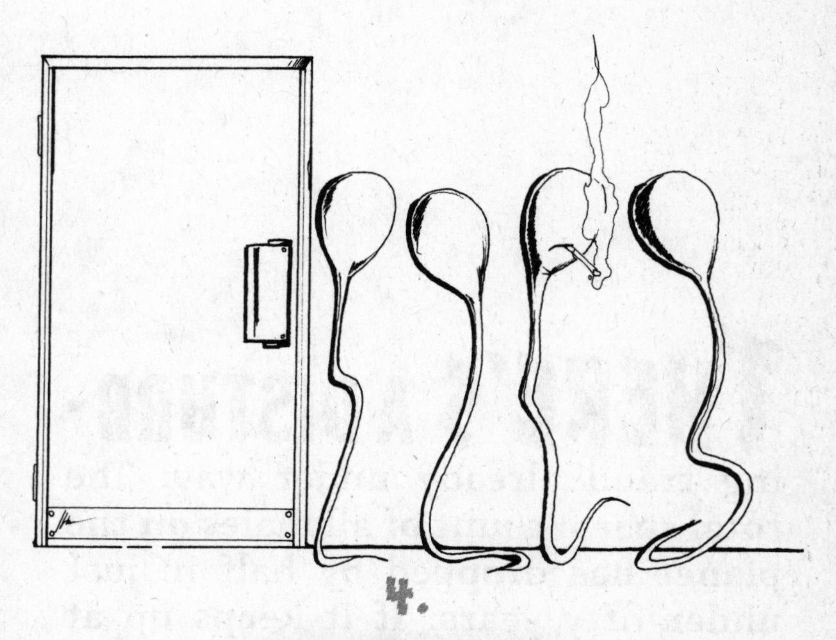

4. STERILITY

THERE'S A DISTURBING trend already under way: The total sperm count of all males on the planet has dropped by half in just under fifty years. If it keeps up at this pace, another fifty years' time may very well see the last human beings born on Earth. It's certainly not the worst apocalypse imaginable; in fact, it's one of the better ones. Sure, there will be the inevitable last-minute, panicked attempt at self-correction as we try to save the species—just like with any other apocalyptic scenario, except that in this case, instead of erecting bunkers to shield us from a nuclear tornado or collecting shotgun shells before the zombie invasion, we'll just try to bone one another hard, fast, and as many times as possible. But it's still the end.

Most evidence of this reduced sperm count comes from citizens of industrialized Western nations, leading many to believe that technology—while inarguably awesome—is nevertheless out to neuter you. Conversely, shitting in a ditch is apparently excellent for fertility. We were first made fully aware of this worrying trend by a Dutch scientist named Niels Skakkebaek, when he conducted a worldwide poll of sperm levels in 1992. Ol' Dirty Skakkeback, as his friends probably called him, went on a veritable world tour of semen, and when he was done—sticky, exhausted, and no doubt walking funny—he found that sperm counts had not only dropped significantly (by the aforementioned half at some estimates), but that even semen with average sperm counts contained a much higher number of deformed sperm than in the past.

This conclusion was soon echoed by other scientists all across the world; scientists like Jarkko Pajarinen, a professor from Helsinki, who conducted a study comparing the testicular tissue of five hundred men from 1981 against men from 1991. He found that the normal sperm production in men from 1981 contained about 56 percent healthy sperm. But by 1991, it had dropped to a little over 26 percent.

Ten years!

Only ten years' difference and the effectiveness of our Littlest Gentlemen had dropped by half! If our collective balls were a company, they'd be filing for bankruptcy. Oh, and one more slightly less frightening, but still embarrassing factoid: Professor Pajarinen also found that the overall weight of the testes had decreased as well. So, long story short, the average modern man has the smallest recorded balls in history. Your verbally abusive stepfather was right! You *are* half the man he is!

But if we're all going infertile, as this data seems to indicate, why does it seem like the planet is becoming ever more crowded with assholes? It's like all the stupidity in the universe collected on the surface of the Earth as retarded condensation. It's true, overpopulation is a problem, but if this fertility trend persists at the current rate, we're dangerously close to being completely sterile as a species. A sperm count of less than 20 million per milliliter is the technical definition of infertile, and at the current rate of progression, that's going to be the average within our lifetimes.

Other Scientific Theories Coined by Your Verbally Abusive Stepfather

- The "Coors Light Makes Your Mom Look Pretty" Hypothesis
- The "Only Faggots Listen to Rap" Theorem
- The "Your Real Dad Was a Pussy and That's Why Your Momma Found Me, Son" Principle

A 2009 study released by scientists from Brunel, Exeter, and Reading universities in England, in conjunction with the Center for Ecology and Hydrology, states that they may have found at least part of the reason for all those blanks loaded into our collective man-clip: water pollution.

Their study found an unusually high amount of chemicals

called anti-androgens in the water supplies tested. Much like holding a purse outside of a dressing room, anti-androgens inhibit your manhood by blocking testosterone receptors, thus lowering fertility rates in males. It's unknown exactly why these chemicals are found in such high volumes in industrialized Western nations in particular, but the working theory is that chemicals from the massive consumer use of pharmaceuticals are starting to enter the water supply through our waste. In other words, prescription meds are being passed en masse into the water supply through urination.

You're literally peeing infertility.

You can see some proof of this happening already by observing the fish living in the water in the most affected areas. They're so severely impacted by the concentrated dosages in their environment that the drugs are actually feminizing the male fish, in some cases causing them to spontaneously change sex. Now, fish and men are entirely different animals, so in no way should this information be summed up like this:

The average person has anti-baby pee that turns dudes into ladies.

That's just a staggering oversimplification based on a very limited set of data. So maybe you shouldn't worry yourselves about it. Even if it is *happening right now, as you read this!*

Jokes Terrible Stand-Up Comedians Would Tell About "Feminized" Fish

- "Yeah! Apparently, it seems they found this information when all the fish started crashing their cars!"
- "What do you call a Zebra Fish with two black eyes? Nothin'! You already told it twice."
- "Have you heard about this? Male fish turning female? They call 'em TransvesTilapia."

But for there to be some real danger of a humanity-erasing plague of sterility, it would probably have to strike both men *and* women drastically.

Which it is. Come on! Would it be in this book if it wasn't terrifying?

The UCLA School of Public Health has found some early evidence that perfluorinated compounds, or PFCs, could be associated with increasing infertility in women. And, though they sound exotic and rare, PFCs are used in pretty much everything: Plastics, pesticides, clothing, makeup—odds are you're wearing or touching something chock-full of perfluorinated chemicals as we speak. The study says that women with higher levels of PFCs in their blood take longer to become pregnant than women with lower levels, if they can become pregnant *at all.* Because more manufactured materials are used and discarded in those pesky industrialized Western nations again, of course they're the ones getting hit the hardest. And right in the babymaker.

There's research to suggest that the fertility of both sexes is in decline due to somewhat mysterious circumstances, but not all sterility-inducing factors are unidentifiable. For example, right now a new crop of genetically modified corn is being harvested. This mutated corn isn't any larger than normal, doesn't have a longer shelf life, and isn't more resistant to disease: It has only one purpose, and that is to serve as a contraceptive.

Seriously.

It's condom corn. It's birth-control maize. The crop is harvested and distilled into a gel that acts as a spermicide, but it could render males who eat it infertile as well. One can only assume that the corn is also quite literally the most de-

lectable substance on Earth, because though ordinary corn bread is undeniably delicious, it's not quite "I don't want to have kids ever again" delicious.

A company in San Diego called Epicyte is responsible for this terrible, terrible idea. They've accomplished this extremely unsettling feat by using a recently discovered and exceedingly rare class of human antibodies that attack sperm. When they found these antibodies—which I should remind you are described as "attacking" something *inside your balls*—Epicyte decided to take the road less traveled and, instead of taking the logical course and killing them with fire, they opted to splice them into corn crops.

Other Epicyte Inventions

- Black Death by Chocolate Cake
- Coke Ebola
- AIDSburgers

But I digress. Epicyte isn't pure evil; they actually want to help. The general idea Epicyte is working on is to create a hormone-free contraceptive that places reproductive responsibility on both men and women equally, rather than sticking with the status quo, which is asking women to take daily insanity pills so sex can be fun again. Epicyte has also isolated an antiherpes antibody and spliced that gene into the corn as well. So their product will not only serve as a sexual lubricant, but also as a contraceptive and an STD inhibitor. Epicyte president Mitch Hein explains how it works in bizarrely dance-centric terms:

> Essentially, the antibodies are attracted to surface receptors on the sperm. They latch on and make each sperm so heavy it cannot move forward. It just shakes about as if it was doing the lambada.

Aside from his strange obsession with the Forbidden Dance, that's an accurate description. The antibodies don't kill sperm, they just render it inactive. It's not like you accidentally eat the wrong corn dog once and so can never have babies again—the antibodies actually have to be continually present to function. If you stop eating the corn, the effect will gradually fade.

And before you start thinking that's comforting, let's revisit what we learned from the genetically modified foods section: Scientists have proven, over and over again, that it is practically impossible to fully contain GM crops. They *will* escape and, what's worse, could actually become the dominant strain of a crop through purely natural reproduction, not to mention the possibility of crosspollinating into other related plants. When you consider that corn is the single largest crop on the planet—sustaining not only our own food, but that of our livestock and, thanks to ethanol, even our vehicles—that's a pretty big field to contaminate. Since it's not an "if" but a "when" the contraceptive corn escapes, that means there is a distinct possibility that the largest crop on the planet will eventually render you infertile if you eat it, forcing you to choose between food or babies.

Food is delicious and babies are loud. And if the question is "Would you rather have a Coke or a kid? A sandwich or a lifelong commitment?" we all know the answer most men would choose. (Hint: It's the dooming-humanity one.)

Not enough examples to worry you yet? OK, one more: There's a popular theory right now that obesity, conventionally blamed on too much pie and couches, could actually be caused by a virus. A fat virus.

It started twenty years ago, when an Indian scientist named Nikhil Dhurandhar found something odd: An epidemic had struck the local chicken population, killing thousands of birds and leaving behind giant, fatty corpses. Eventually a bizarre type of adenovirus was found to be causing the deaths and now, twenty years after the chickenpocalypse, it's happening again.

In humans.

Another strain of the same disease, called AD-36, is being found in increasing numbers in human fat tissue. And, as Professor Richard Atkinson of the University of Wisconsin found in a related study, it does have the same obesity effect in humans as it did in chickens. He tested five hundred people for the AD-36 strain and found that those infected by the virus weighed noticeably more than the uninfected. Even after isolating and destroying it in patients with antiviral drugs, the virus' fat-making effects were not reversed. So, like a can of viral Pringles: Once you pop, you can't stop (being fat).

It's not just limited to this one bizarre strain, either. There are several other pathogens linked to obesity in the animal world, and any one could make the jump just like AD-36. Of course, this is all in a chapter about sterility, so let's get to the matter at hand: Let's assume there's a fat plague ravaging the world and that eventually everybody will end up big boned and burger laden. Humanity still has urges, and what is deemed attractive in the oppisite sex can be quite flexible. So we're having fat, sloppy, roll-slapping sex, so what? So it's not getting us anywhere, that's what. A study at the Academic Medical Center in Amsterdam tested three thousand women struggling with fertility problems and found that chances of successful pregnancy reduced by a staggering 4 percent with every additional Body Mass Unit: The more obese the woman, the less her chance of pregnancy.

> This is known as the No-Fat-Chicks Theory of Evolution, and is currently being espoused by both Dutch fertility scientists and the bumpers of pickup trucks with gun racks everywhere.

That means that there's a virus that inflicts irreversible obesity that in turn renders us infertile, not to mention the fact that our water is "feminizing" all the males, overall ball size has shrunken within a generation, and cornflakes turn your sperm into all-night dance machines that shake, shake, shake it until they die. It's not hard to see that this is actually one of the most likely doomsday scenarios threatening our species today. So if I were you, I'd start fucking right now.

We're going to need the head start.

5.

NEW ENERGY

ASIDE FROM BEING less profitable and more difficult to implement than conventional methods, so-called green energy has another, far more important problem to overcome: It's for pussies.

Or at least that's the popular consensus. Sorry, hippies, I want to save the world as much as the next Bruce Willis, but there's simply nothing sexy about our available alternative energy sources. All of our past major fuels have had at least one important thing in common that eco-friendly power lacks: They could fucking kill you. Fossil fuels burn, nuclear power irradiates, and coal once killed a man in Kentucky just to win a bet. Even pre–Industrial Revolution power was distilled from pure badass. Whale oil was the fuel of choice, and this was at a time when whales were poorly understood leviathans. They were quite literally demons of the deep—near legendary creatures the size of your entire goddamn boat—and the only way you could read your copy of *Pride and Prejudice* after sunset was to slay that sea monster with a fucking spear and render its fat to light your lamps. But solar panels, windmills, hydrogen fuel cells—shit, you might as well power your car on kitten hugs.

Well, lucky for the planet, science is about to change all that and make green energy as sexy and dangerous as she is renewable and clean. It's just that it may be at the expense of all of our lives.

THE FUTURE OF HYDROGEN ENERGY

At the Rutherford Appleton Laboratory in Oxfordshire, England, a group of scientists is refining designs for a new energy facility they call HiPER (high power laser energy research). This facility, apart from proving that scientists cheat at acronyms (if you're being fair about it, it should be HPLER, but "Hipler" apparently isn't sexy enough, sounding more like

Adolf's rap alias than a high-tech research center), may also solve the world's growing energy crisis in the most awesome way that science knows how . . .

With lasers!

The HiPER facility is pursuing the heretofore presumed pipe dream of nuclear fusion, which they hope to achieve by using a combination of hydrogen and superlasers. The idea is to build a superefficient, giant, rapid-fire laser—not for world domination or eliminating that bastard James Bond, but to superheat a tiny little pellet of hydrogen. The hydrogen would be dropped down a 33-foot concrete and lithium shaft, where, upon reaching the dead center of the reaction chamber, it would be struck with 192 separate lasers, the combined output of which would be 500 *trillion* watts (or one thousand times the power of the entire U.S. national electricity grid) delivered all at once. This creates temperatures of more than 100 million degrees Celsius in the hydrogen pellet, thus replicating the same fusion process that powers our sun. Tubes of water surrounding the chamber superheat from the reaction, the steam is converted into electricity, and hydrogen pellets learn never to cross mankind again.

You see that, hippies? That's how you fucking do green energy. Maybe some people won't trade horsepower and performance for mileage, but even if it tops out at 16 mph, comes in only bright pink, and the engine sounds like Tiny Tim singing Hello Kitty songs, there's still not a man alive who wouldn't drive an electric car powered by goddamn *laser fusion*.

Manlier Ways to Power a Vehicle

- Whiskey
- Punches
- Pulling it with your dick

All right, so maybe that's not strictly true: The laser fusion would be converted to electricity in a separate station rather than the actual car itself, but hell, even lasers by proxy is more lasers than you have right now. Let me throw a little bit of complicated math at you to explain why that's a beneficial development:

If "lasers" = "good," then "more lasers" = "gooder."

Clearly, the logic is infallible.

Though this technology isn't quite feasible yet, it is close: Technicians at the National Ignition Facility, part of the Lawrence Livermore National Laboratory in California, while typically busy having the most badass job titles in the history of everything, are also spending their time carrying out early versions of the experiment right now, and eventually hope to combine their efforts with the new HiPER facility to create a viable electrical grid.

A few worrisome things about this process, however, are the temperatures and pressures created: Akin to a tiny, controlled sun, remember? So, how much of an issue is containment? The process has already been proven viable by the National Ignition Facility, and they're planning to scale up to a larger form when the HiPER facility is built, yet little is said of containing this energy in the event of a failure. Forgive me for being a tad bit worried when somebody borrows a plot point from a comic book supervillain's best-laid revenge schemes just to power their home, but creating a miniature sun on the surface of the Earth seems like the God King of bad ideas. After all, back in the '40s there were rumblings that

Worse Ideas Than Creating a Miniature Sun on Earth

[Error: no data]

the first nuclear tests would ignite the oxygen in the atmosphere, leading to a global chain reaction that would, in turn, ignite all of the air on Earth—and they still went ahead with testing the bombs.

I hope this kind of thing helps you see why trust is still an issue, Science. Seeing as how you risked lighting our lungs on fire just to build a better bomb, I don't know if we as a people are ready to let you penetrate our pristine, virginal electrical grids with your hot, throbbing lasers. It might just be too soon to go all the way with you; at least prove you're not going to strangle us for kicks afterward.

THE FUTURE OF WIND ENERGY

Even assuming that the fusion lasers don't get us, there's also a new experiment in alternative power under way in Canada. It's called the AVE, which is short for Atmospheric Vortex Engine (Note: That's how you do acronyms, scientists. The pseudobiblical tones succeed in scaring the shit out of us with three little letters alone), and it's got the scientific community in a whirl.

Well, more like a whirlwind.

Actually, exactly like a whirlwind.

That's what the AVE is, after all: a giant, artificial tornado, reaching up to ten kilometers straight into the atmosphere. Like slapping a leash on a volcano, the AVE proposes to literally tame a natural disaster by creating and controlling its point of origin, then harnessing the resultant power. The principles it functions on are pretty straightforward: Tornados are formed by cool and warm air mixing in sufficient quantities. So it's really just a simple matter of forcing the warm air upward far enough to contact the cool air above . . . and then attempting to harness a force of nature that has killed human beings and destroyed cities since the beginning of time.

Listen: Canadians are great—national health care, clean cities, a very polite disposition—but there's a limit to how much faith Americans can have in their neighbors up north, and literally reaping the whirlwind might be a little beyond that.

Assuming they don't immediately bring forth the wrath of God for usurping his powers, a ring of turbines on the ground would both maintain ideal tornado conditions (perpetually pumping warm air in a circular motion) and also do double duty in harvesting all that energy, just like windmills. The AVE generating facility would be a low, squat building surrounding a giant cylindrical tube that's open to the sky. The artificial tornado would emanate from this tube, with only the lower part contained by the structure. Essentially, the large, flat ring and containment wall would look like the stand of a trophy, with the trophy itself being a several-kilometers-high twister. If it helps you to picture it, just think of it as a giant award for World's Scariest Building.

Things Canadians Are Good At

- Politeness
- Hockey
- Syrup

Things Canadians Are Bad At

- Basketball
- "Yo momma" jokes
- Starting wars

Things Canadians Are Questionable At

- Wrangling tornadoes

Its inventor, Canadian scientist Louis Michaud, says the AVE would also serve to stabilize the localized weather around it—maybe even facilitating rainfall by assisting the transportation of ground moisture to cloud level.

Win, win, and win, right? Farmers get drought protection, the world gets energy, and metalheads get a new thing to watch instead of laser light shows!

At its heart, it's the same principle as conventional wind farms, and those seem pretty quaint and harmless, right? The chief difference here is that conventional wind farms consist of giant pinwheels in rolling green fields capturing the essence of a gentle summer breeze, while the AVE is a gargantuan black tower in the center of a dry lakebed in Utah topped by a giant, artificial, eternal typhoon. You read correctly. This isn't mere theory. There is actually a working prototype of the AVE. It was built solely as a proof of concept, so the whirlwind it generates is not kilometers high and the energy it reaps is insignificant, but it does work! The rest of it is just a matter of scale now. Oh, and just to mix things up a bit, the experimental AVE tower is also used to generate something called Fire Spirals, presumably just in case Harry Potter tries to meddle in your whole "eternal whirlwind" plan.

Other Potential Uses for the AVE Tower

- Collecting storm data in a controlled environment
- X-treme kite testing
- Murdering hang gliders

The most likely locations for a finished tower will be in the temperate northern hemisphere (like the northern United States, southern Canada, and most Western European nations), because the height of the troposphere there drops to a measly seven kilometers. Provided that a tornado can be prompted to reach that high, the natural difference in atmospheric pressures between the anchor point and top of the twister should help to maintain the vortex. So really, if you just have a large enough base unit installed somewhere in the north that supplies the occasional burst of circulated hot air, you should have a fully functioning domesticated whirlwind. One that is theoretically kept fully in check behind its tiny containment wall, despite the funnel rising ten kilometers into the sky into space.

Uh . . . *theoretically*.

Nobody's ever actually, y'know, made a tornado their bitch before, and anything from a high wind to a warm spell might *theoretically* cause the twister to jump the wall and go rampaging away—free to wreak a terrible vengeance upon mankind for their arrogant attempts to confine it. Louis Michaud is an engineer, not a meteorologist, so just to be on the safe side he recommends that the AVE be built away from populated areas—as he says, "just in case." And when a man proposes to build a machine that creates tornados, "just in case" is the last thing you want to hear from him. That shit would worry Dr. Doom. There is simply no place for a "just in case" when you consider that the AVE is not actually a self-sustaining power plant; it's an *addition* to a power plant. The warm air the AVE needs would be cast off from already existing power stations, so the idea is to supply every power plant in the world with an Atmospheric Vortex Engine . . . including the nuclear ones.

Especially the nuclear ones.

And if you think the prospect of a nuclear tornado in every major city in the world doesn't sound like an "end of the world" scenario to you . . . well, good for you, Batman. Put down the fucking book and go do something useful, like fight crime. Your giant, fearless balls are wasted here.

THE FUTURE OF SOLAR POWER

Solar power reaps the very rays of the mighty sun, but it does it too passively. You just lay out a panel and let it get warm. That's boring and, unless you throw some tits and a bikini on said panel, pretty unsexy. Well, no longer! Thanks to space-based solar power systems—or, as their friends call them, Barely Controlled Orbital Death Rays—the future of solar power is as awesome as it is insane. Everything else about the SBSP is like duct-taping Awesome to the back of Badass and then deep-frying it all in some "Fuck Yeah." SBSP refers to a satellite in permanent orbit that basically collects energy from the sun and then fires it back to Earth in the form of *giant lasers from outer space*.

"Giant Lasers from Outer Space" Sounds Like:

- A B-grade Japanese monster movie
- A folk-rock circus troupe
- A Talking Heads album
- A frightening prospect

The chief advantage of SBSP is that there's no diurnal rotation to worry about—no alternating periods of day and night that effectively cut the energy-gathering time in half, because it's always sunny in space. And that's great! That's just the kind of optimism you don't often hear when referring to lasers fired from an irradiated, vacuous void. Way to think positive, SBSP!

An American company named Solaren says they have a plan for a fully functioning SBSP station with available technology, and they're not the only ones: A private Indian corporation has thrown its hat into the ring, as has the Japanese Aerospace Exploration Agency (JAXA). Any number of entities expect to have working prototypes circling the Earth in the near future, firing buckets of sweet, hot, lasery goodness into the mouths of naughty, subservient little power stations all across the globe. And while these stations will be strictly limited to remote locations like deserts and mountaintops at first (which is pretty much like saying "just in case" again), the SBSP companies do expect to expand the operations globally. The JAXA in particular sees a future where power needs would be accessed "along much the same lines as a cell phone call," and you'd simply "put in a request" to have power harvested from the sun itself fired at your exact location from orbital lasers in outer space, thus enabling you to either jump-start your dead car battery or hold an entire city for ransom, depending on your needs and moral flexibility. So if you were concerned before about stuff like radiation from your cell phone, or living beneath power lines, you might want to start subscribing to *Bunkers and Canned Food Monthly*, because unfortunately, that homeless guy in front of the library was right: Soon we actually might all be softly microwaved from the inside out by Japanese space lasers.

But the JAXA's microwave lasers aren't even the scariest of the planned systems, and that's why we're mainlining some Solaren. As mentioned, Solaren is a California-based power company that hopes to launch solar panels into space. What was not mentioned previously is that these solar panels are more than a kilometer wide, and would ideally channel hundreds if not thousands of megawatts of energy once operational. The principal electric company of the state of California, the Pacific Gas and Electric Company, in a sign of either optimism or proof that they've given in to space-based solar terrorist demands, has agreed to purchase all of this energy if Solaren does succeed. So Solaren already has a buyer; they just need to build the product, which they expect to have completed by 2016. All told, the entire system would annually generate up to 4.8 gigawatts of power in geosynchronous orbit, convert it to radio waves, and then fire it to a collecting station on the ground. Because the radio waves would be spread into a rather wide beam, the hope is that they wouldn't be too dangerous to everything below them.

Oh, but that whole "harmless to everything below" thing is a far cry from Solaren's initial goal: Absolute control over the weather for destructive purposes.

Situations Where the Word "Hope" Is Comforting

- Church services
- Funerals
- Elections

Situations Where the Word "Hope" Is Not Comforting

- Describing the certainty you have that vast areas of U.S. soil will be immune to the lasers you are going to fire at them.

What? That came out of left field, didn't it? Surely it can't be true: Why would California be buying sci-fi power stations from a space weapons developer? It's almost too crazy to believe, and if it weren't California—home of the bad decision—you probably shouldn't. But in this case it's true. The inventors, Jim Rogers and Gary Spirnak, wrote in their patent application:

> The present invention relates to space-based power systems and, more particularly, to altering weather elements, such as hurricanes or forming hurricanes, using energy generated by a space-based power system.

The theory is that by either heating up or diverting heat from a burgeoning hurricane, Solaren's panels could either destroy or encourage the infant storm. If you heat certain parts of a tropical storm, thus lowering or raising the pressure differential between the upper and lower layers of a hurricane, it is possible to weaken, strengthen, or completely dispel it, depending on your purposes. Let's repeat those points for emphasis: Kilometer-wide solar satellites firing lasers that destroy or create hurricanes.

That shit is being built right now, destined for the skies above California, and it will be operational by 2016.

And you promised to pay them, California. You know what happens to people who don't pay their debts to shady loaners? They get their legs broken. Your loan shark has a hurricane laser, and everybody is well aware that you're broke as hell.

This begs an interesting question: Can an entire state go into the Witness Protection Program?

NATURAL DISASTERS

Nobody kills like Mother Nature. She is the undisputed master of terror — inventor of the spider and the giant squid, producer of the parasite and the flesh-eating bacteria. If Gaia is the source of all life, she's also the cause of all death. And while death is a natural, necessary part of the circle of life, some of her methods seem like downright overkill; if we're all going to die of old age, having the Earth explode and then drown itself in fire just seems a bit mean spirited, doesn't it?

But hey, love her or leave her, you owe your life to Mother Nature, and you live all of that life on her very body. And judging by the following examples, she really fucking hates you for it.

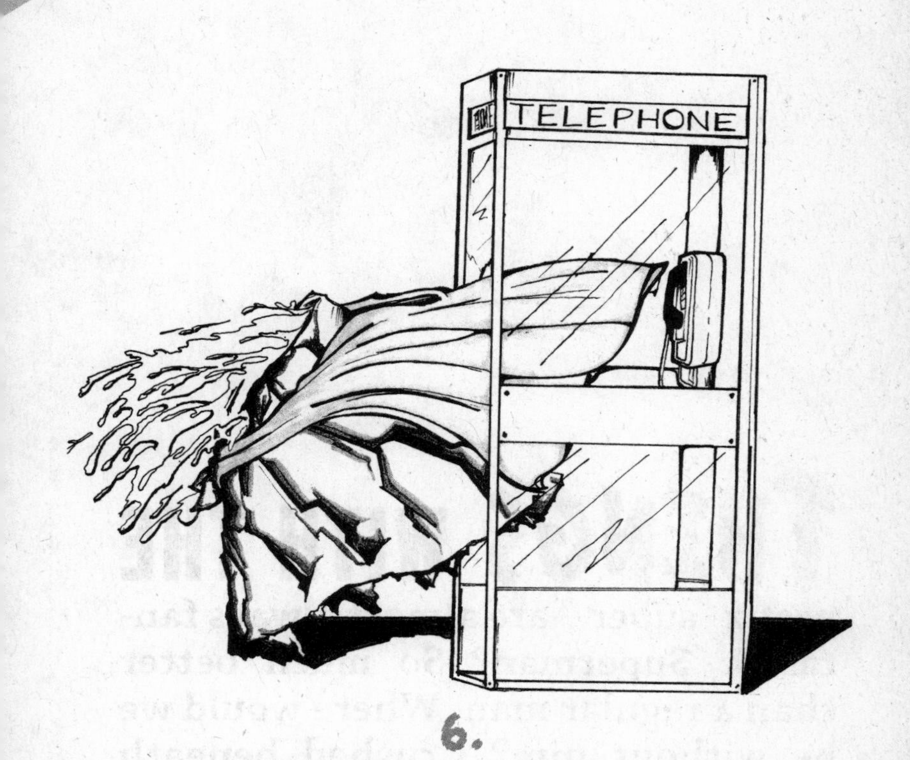

6.

SUPERVOLCANO

THINGS WITH THE

prefix "super-" are almost always fantastic: Superman? So much better than a regular man. Where would we be without him? Crushed beneath the heel of Lex Luthor's Angro-bots, that's where. Stunt comedian Super Dave? Just like a regular Dave, but five times more hilarious. Supersize it? Fuck yes! What do I look like, a guy who eats regular-size meals? You add "super" before a word and it's like marketing crack. Super anything is great!

Just all around fantastic.

Really.

No exceptions.

Well, maybe one exception: the supervolcano. That's not so great. If you were hoping it was "super" like Superman—a regular volcano given superpowers to protect humanity—it's not. But it *is* sort of like supersizing: It's a volcanic eruption so big *the whole world* can share! Finally, something that touches all of mankind!

. . . and turns them into *screaming ash*.

A supervolcano occurs when magma builds up below the crust of the Earth, but can't quite break through. All the heat, the gas, and the pressure—it all keeps building up until the Earth just can't take the pressure anymore and bursts. So to sum up: A typical volcanic reaction is like a normal person throwing a fit—a little eruption just to vent the pressure, but generally keeping the devastation to reasonable levels. But sometimes the planet just holds all that fury inside until it snaps. Except by "fury," I mean burning rock, and by "snaps," I mean *superexplodes*.

There have been only a handful of these supervolcanoes in all of history, but just those few have been responsible for mass extinctions, global weather changes, and sometimes even small ice ages. Supervolcanoes must, at the minimum, consist of at least 1,000 cubic kilometers of magma. That's basically a small country's worth of material, and it's all *lit on fire and flung through the air*. The eruption would trigger massive earthquakes, the lava would burn through everything for thousands of miles around, and the ash would choke out the light from the sky. They even keep destroying after they stop: Supervolcanoes don't leave cones like a normal volcano; they create massive calderas more akin to an impact crater, because so much mass is ejected that the Earth simply collapses around it.

Maybe it can be of some small comfort to you if you consider that the last supervolcano was a long, long time ago. Why, over twenty-six thousand years ago as a matter of fact! I can barely remember starting to write this sentence, so twenty-six thousand years is a lot longer than I can even comprehend. And if you're anything like me, that's enough time to make you feel safe—shielded by the buffer of history. It matters little that the second-to-most-recent supervolcano, over seventy-five thousand years ago at Lake Toba in Indonesia, caused a volcanic winter, triggering an ice age that lasted for more than a thousand years, killed off between 70 and 90 percent of the human race (depending on which estimate you

How to Deal with a Supervolcano

- Duck and cover
- Sit and spin
- Bump and grind
- Whatever else distracts you for a brief moment from the burning inferno that you now call home

use), and formed sulfuric acid in the fucking atmosphere (you know, that thing you breathe in, and live inside of? That was acid). Hey, just as long as it's not happening right now, you shouldn't have to worry about history. Because let's face it, if you paid attention to history you'd never leave the house; that shit is terrifying.

Wait, sorry, I'm getting an imaginary note passed to me here.

One second while I pretend to read this . . .

Oh, awesome! Says here that a supervolcano could actually be about to erupt right now—*right fucking now, right in the United States of America.*

Historians call this the "Ostrich Defense" and some observers* note that this method has a 100 percent success rate.

*NOTE: Observers are hungry lions.

You might know ground zero: Yellowstone National Park, home of postcards that your grandparents send you, scenic vistas, Old Faithful, and, apparently, terror. The single most potentially destructive volcano on Earth, the Yellowstone Caldera in Wyoming, is now showing strong signs of becoming active again. It's not only a proven supervolcano—Yellowstone has had previous supervolcanic eruptions, 2.1 million, 1.3 million, and 640,000 years ago—but it's also a "geothermal hot spot." Supervolcanoes and geothermal hot spots are like a disastrous peanut butter and a devastating jelly: The two don't always go together, but they're exponentially improved when they do. The hot spot beneath Yellowstone refers to the end of a gargantuan plume of magma, the molten rock that swirls around in the Earth's mantle below the solid rock of the surface. Under ideal circumstances, that plume would peter out about fifty miles underground, but not in this case: Just the tip of this mantle plume is several miles wide, and it's been sitting down there for thousands of years, slowly melting the underside of the Earth's surface away, until it has eventually encroached to within a few hundred meters of ground level. But while the tip of this mantle plume is scary enough, you actually have to worry about the entire thing if it bursts. Just like an overinsistent teenager, what starts with "just the tip" will inevitably end with a full-on shaft. If any part of the plume breeches, the vast pressures beneath will force all of it out. And all there is right now is the thinnest veneer, a sheer G-string of dirt, really, that's keeping that entire hot, smoking shaft of fiery death from spurting all over the Earth like the devil's money shot.

The chief indicator that the Yellowstone Caldera might be becoming active again comes in the form of a recent "swarm" (worryingly enough, that's actually the official term) of earthquakes registered there. At the end of 2008 there was a period of rapid-fire, low-level tremors in and around the Caldera—around eight hundred separate earthquakes in just under a week, which, if you're counting, is 799 more earthquakes than it takes to scare the shit out of everybody under the best of circumstances, much less when they're emanating from a giant organic time bomb like the Caldera. Robert B. Smith, an emeritus research professor of geology geophysics at the University of Utah, believes the unusually high earthquake activity could be a sign that the volcano is reawakening. Or, as Professor Smith himself puts it:

Supervolcano Porn Titles

- *Steamy Eruptions*
- *Sizzling Encounters*
- *Spurting Plumes*
- *Burning Love 3: Literal Edition*

Aaaaaaaaaaaaaaaaaaaaaaah fuck fuck I wish I could run but there will be fire eeeverywheeere ahhhhhh.

Well, he was probably thinking that, anyway.

Considering that the minimum size for an eruption to be considered "supervolcanic" is 1,000 cubic kilometers, and the pool of magma beneath the Yellowstone Caldera is estimated to be 28 by 45 miles across, screaming panic seems like the most logical reaction. Now, I'm not exactly sure what those numbers translate to in kilometers, because I use God's System of Measurement (which is pounds or ounces or, really, whatever we make up off-the-cuff here in America), but I'm pretty sure that's equal to eight bazillion cubic kilometers of magma. And that, my friends, is eight bazillion more kilometers of flaming rock than anybody should be comfortable with.

Because volcanology is not an exact science, experts have little to no idea of what to expect from an active supervolcano. They believe that four signs, like metaphorical horsemen to the Rock Apocalypse (which would be the best metal band name ever) will herald its eruption. First, the ground will rise from the pressure of all that magma, then geyser activity will increase, swarms of earthquakes will register, and a large release of volcanic gases will occur before just before the eruption.

Database Error: Fart joke not found. Please insert Taco Bell reference for humor substitute.

So far, *three of these four signs* are present in the Yellowstone Caldera! It's been named a High Threat for Volcanic Eruption by the U.S. Geological Survey, who went on record as stating that the eruption from Yellowstone would entail "global consequences that are beyond human experience and impossible to anticipate fully."

That is without question the single most ominous quote ever to be issued from a government agency, and that's coming from the Geological Survey Team! That's the least threatening team that has ever been assembled short of the Super Friends, and if they're issuing quotes so ominously epic that they're almost biblical, well, I don't want to say it's time to panic . . . because that time probably passed about a year ago. This is more like "make your peace" time, if anything.

Volcanologists do say that, thanks to their advanced technology and years of study, they can give us *something*: About a week's time to prepare. While that may seem an inadequate amount of time to try to figure out how to run away from an entire planet, at least you have a whole week to live in perpetual fear! An entire week! Why, that's enough time to start a garden! That's enough time to get over the flu! That's enough time to buy and receive something from eBay! And hell, if you're really lucky, that might be enough time to practice putting matches out all over your body in order to help brace yourself for the coming storms of fire that will consume all of your flesh, turn the air to poison, and kick off a nigh-eternal winter!

. . .

Or you could knit a scarf!

7.

MEGATSUNAMI

THE JAPANESE CALL it *iminami*, which means "the purifying wave." They do this partly because it is a wave of such devastating strength that it completely erases the land of all impurities (impurities, in this case, being such blighting defects as your house, your car, and probably you, depending on how fast you can run and how well you float). But they also do this because they are much, much better at elegantly naming horrific events than the English-speaking world. We have a name for it too; we call it the megatsunami. Judging by the American tendency to just slap superlatives on existing terms, I guess we should just consider ourselves lucky that it's not called the Biggie Wave or the Supersize Water Punch.

Other Examples of the American Tendency to Add Superlatives to Existing Terms Rather Than Create New Titles for Epic Disasters

- Supervolcano
- Hypercane
- Megaquake
- Überdiarrhea

The concept behind the megatsunami is simple: If you throw a pebble into the water, you'll see a reaction in the form of a rippling wave. If you threw 500 million tons of rock into the water, you'd be a total dick, but you would also see a wave of proportionate size . . . one so powerful that it jumps forests, snaps cities in half, and floods entire coastlines. But the phenomenon was predominantly thought to be a myth until just recently. See, scientists already know how tsunamis are triggered: Earthquakes, volcanic eruptions, and other seismic events create waves that can crest at tens of meters high and hundreds of kilometers long that travel vast distances, surging onto land with unstoppable force. Conventional tsunamis, however, don't really look like waves; they're more akin to gargantuan tides, and the damage they do, though terrible, is mostly through flooding and not so much due to the impact of the water itself.

But in 1953 in Lituya Bay, Alaska, geologists searching for oil stumbled across something much, much worse. By taking measurements of the tree line along the coast, they came to realize that a cataclysmic wave had completely destroyed the area in recent history. Seeing as how the bay was mostly isolated from the open ocean, they were able to determine that a gargantuan landslide was the likely cause. Forty million tons of debris had to tumble into that bay in order to spawn a tsunami large enough to account for the destruction they were witnessing. This would be a wave unprecedented in recorded history. A wave with an initial surge height estimated at over 1,700 feet.

Armed with this dire new information about a terrifying and impending threat, the geologists decided to issue absolutely no statement whatsoever, addressed to nobody, which would have probably just read "fuck it," if they had even bothered to give it to anyone in the first place.

As the geologists in question whiled away their time—presumably playing a few games of grab ass and maybe frolicking hand-in-hand through a sun-kissed meadow—Lituya Bay was busy preparing another watery jump kick to the throat of reason. And only five years later, in that exact same bay, it happened again. An earthquake that measured 7.7 on the Richter scale caused a chunk of the Lituya Glacier to drop three thousand feet into the bay waters below. After the initial surge that topped out at 1,700 feet (that's taller than the Empire State Building), a much more modest, practically meager wave with an initial height of only 1,000 feet swept

Excerpt from Alaskan Geologist's Log, Circa 1953

"My God, I've discovered evidence of an entirely new scale of disaster!"

"We must tell the world!"

"That sounds hard . . ."

"You're right. Screw it. Wanna beer?"

"That also sounds hard. Will you pour it into my mouth for me?"

across the bay, and out to sea. A local fisherman, Howard Ulrich, and his son were not only caught up in the ensuing megatsunami, but even managed to survive it—presumably by virtue of their giant, grizzly bear–sized balls and maybe some sort of Eskimo Magic. They reported being carried just behind the crest of the wave, which surged dozens of meters above the bayside cliffs, and *over* the local forest . . . while still in their fishing boat! Another survivor, Bill Swanson, gave this description to the local papers:

> The glacier had risen in the air and moved forward so it was in sight. It must have risen several hundred feet. I don't mean it was just hanging in the air. It seems to be solid, but it was jumping and shaking like crazy. Big chunks of ice were falling off the face of it and down into the water.

When asked to describe what happened next, Swanson says he is largely unsure, because "the wave started for us right after that and I was too busy to tell what else was happening up there." While it normally might be safe to assume that Mr. Swanson was "busy" futilely sobbing in the fetal position and cursing the wicked God that unleashes such horrors upon the world, one must keep in mind that Bill Swanson was

Pressing Questions Raised by This Quote

What does that mean? Not hanging in the air solid but rising several hundred feet? How did it raise itself above the mountain? Did it grow temporarily? Was it raising its hackles, like some sort of giant, angry, mountainous dog? How can you be so calm?! THE DOG MOUNTAIN SHOT A SKYSCRAPER OF WATER AT YOU!

an Alaskan native, and Alaska in the 1950s was basically a nigh-unsurvivable land of extreme temperature, severe terrain, and all-night grizzly bear mauling orgies. So in this context, it's pretty safe to assume that Bill Swanson's laconic statement that he was "too busy" to properly witness the largest wave in recorded history means he was probably rabbit-punching a Sasquatch in the stomach because it owed him money.

But surely this particular phenomenon is too localized and too awful to occur anywhere else in the world but the harsh land of Lumberjack Wrestling and Salmon Cola?

Of course it is! Breathe a sigh of relief, friend!

Now hold that breath for the rest of your life, because that was a filthy lie. These things happen all the goddamn time:

- Eight thousand years ago, a rockslide on Sicily's Mount Etna caused a megatsunami that swamped three continents.
- Eight thousand years ago, one of the world's largest known rockslides caused a megatsunami originating in the Norwegian Sea. Scientists have found sediment from the slide as high as 65 feet above sea level, and as far as 50 miles inland . . . in fucking Scotland!

Though these intervals sound like vast periods of time, geologically speaking they're like blinks of an eye. Nature has megatsunamis like you would change the channel—it's just not a big deal anymore. But as human beings, it's hard to see much beyond our own lifetimes, and events thousands of years ago just cannot be conceived of as threatening. So let's skip right forward to another megatsunami in recent history.

In northeastern Italy in 1963, a landslide into a lake above the Vajont Dam triggered a small, localized megatsunami that completely destroyed five nearby villages, killing two thousand people (some sources say more). The dam, one of the highest in the world, stood 860 feet high and towered directly above the town of Longarone. When 270 million cubic meters of earth collapsed into the waters at a speed of roughly 70 mph, the resulting megatsunami was said to be over 800 feet high. This, when combined with the already massive height of the dam itself, led to a wall of water crashing down upon the quaint village of Longarone from nearly a quarter mile in the air. It is said that due to the particular location of Longarone—sandwiched between two towering cliffs, with the dam at the far end of the valley—if you stood in the town and faced the wave at its peak, it would have blocked out the entire eastern sky. It is also said that if you stood facing the wave at its peak height, you were both entirely fucked and totally dead, and therefore quite unlikely to tell anybody about this whole "blocking out the sky" thing in the first place, so you might want to take that story with a grain of salt.

Though that knowledge certainly takes just a little of the joy out of life,

Things to Do While Waiting for a Quarter Mile of Water to Crash Down on You

1. Pray.
2. Cry.
3. Get a head start on dog-paddling.
4. Jump in the air and flap your arms on the off chance you have secretly had the power of flight all this time and were not truly motivated to use it until now.
5. Complete 1/16 of a crossword puzzle.

it's not exactly world threatening. An Alaskan fishing crew here, a quaint Italian village there; it's probably nothing that can affect you, right?

Your optimism is so endearing!

But no, at some point in the near future there *will* be a megatsunami—one so massive that it will leave entire continents drowned in its wake. Because right now, off the northwestern coast of Africa in the Canary Islands on the isle of La Palma, there is an entire volcano ready to collapse into the water. A 1949 earthquake split the island's southernmost mountain, Cumbre Vieja, completely in half—opening a fissure that caused the entire shore side of the volcano to shift nearly 7 feet down toward the water. The endangered half of the volcano has an estimated volume of 1.6 million cubic feet, which gives it an approximate mass of 1.5×10^{15} kilograms. Basically, you know you're in fucking trouble when the numbers used to explain how much shit you're in need other, smaller numbers to explain *them*. In short, this fissure puts more than 100 cubic miles of land in danger of sliding into the sea at the next serious volcanic eruption, so it kind of sucks that it's literally the most active volcano in the area. On the upside, this proves conclusively that God has a sense of humor. On the downside, your fear of dying a horrible death is apparently his favorite punch line.

When the mountain falls, the ensuing wave will initially reach heights of more than 2,000 feet, but would likely settle out to a paltry 100 when

More of God's Favorite Punch Lines

- "Liquor? I hardly knew her!"
- "Because he was stuck to the chicken!"
- "I have a wife and kids!"
- "I want to live! I want to liiive!"

it hits land . . . in New York, in Boston, in Florida—the entire eastern seaboard of the United States actually, as well as parts of Brazil, the Caribbean, and Canada. At that point, the wave would be moving at a speed of nearly 700 mph (that's nearly the speed of sound) and with enough force to uproot entire cities like weeds, drag broken skyscrapers miles inland, and generally just erase life like God spilled a bottle of Wite-Out on the Western Hemisphere. The wave would travel nearly thirty miles inland, completely submerging the major population centers of the United States before dragging all debris—human or otherwise—out to sea when it's drawn back. A weakened but still devastating wave would hit across the entire Atlantic seaboard, but the brunt of the impact is squarely on American soil.

The resulting death toll and damage would be devastating to billions of people, and the lasting economic impact would completely destroy the modern way of life, effectively sending everybody not drowned or crushed by waylaid cities directly back to the Stone Age. The new bodies of standing water, millions of corpses, and unsanitary conditions would most likely wipe out the rest of the survivors with disease, but these are Americans we're talking about here, goddamn it, and there's just no way they'll be *totally* wiped out by one little son-of-a-bitch wave!

And that's good, because thanks to the rippling effect, there are going to be ten more right behind it.

8.

HYPERCANE

GLOBAL WARMING

and climate change can cause any number of problems—from food shortages to drought to inclement weather—but they do not always work in such subtle, gradual ways. Changes in the environment don't always function like a cancer, killing you slowly over a long period of time. Sometimes the environment just loses its damn mind, and that's when an event called a hypercane can occur. If the normal consequences of a shifting climate are akin to a metaphorical disease infecting the world, a hypercane's consequences are a metaphorical Bruce Willis: That is to say, if global warming might kill humanity slowly over a period of generations, a hypercane is going to tie a fire hose around the world's neck and then throw it off an exploding skyscraper.

A hypercane is a hurricane on a global scale. With winds up to supersonic speeds, a hypercane doesn't just dismantle and destroy what it touches—it utterly disintegrates it. They can be the size of a continent, the conditions that spawn them could also render them self-sustaining, and their long-term effects are beyond disastrous. In short, it's like a hurricane on death Viagra: In every way it is bigger, stronger, and longer lasting than the worst hurricane you've ever seen.

Though it's a long shot that a hypercane will ever naturally occur again (many theorize that hypercanes have been present for, if not responsible for, some major past extinctions), if it ever does happen, it will be what scientists refer to as a "planet killer"—which, incidentally, is one of the reasons scientists don't get invited to many parties. I'm not saying that these scientists are exaggerating the threat; it's just that there are more sensitive ways to deliver terrible news. Doctors dealing with terminal patients don't break the traumatic news that a patient has cancer by telling them it's "like the atom bomb of diseases"; they don't tell AIDS patients that they have the disease equivalent of "a gun shooting you from the inside out"; and they don't explain leukemia to terminal children by telling them that it's "like the bogeyman lives inside your bones." So scientists are a little tactless, sure, but unfortunately that doesn't make their predictions any less true. While a compassionate soul would tell you that, in the event of a hypercane, we'd all go out peacefully in our sleep, dreaming of past loves and warm summer days, my mother taught me that honesty is the best policy. And in keeping with that philosophy, I should tell you it's far more likely that not only would your skin be sheared off by a supersonic wind, but also it would afterward become a deadly storm-borne projectile that would probably continue on to impale your entire family.

More Horrible Ways to Explain Diseases:

Parkinson's: Like poppin' and lockin' in hell.
Alzheimer's: Like having tiny zombies feeding on your brain.
Herpes: Sex pimples.

It's not like we didn't have any warning, though, as 2008 was one of the most devastating hurricane seasons on record. Massively destructive individual hurricanes like Katrina and Rita in 2005 don't even rank among the ten most powerful storms of all time. But you certainly can't just dismiss this disturbing trend of increasingly stronger storms on the basis that they haven't yet produced the strongest ones ever recorded—that's like telling yourself that the ravenous pack of wolves following you for the past fifteen minutes are nothing to worry about because you saw a much bigger lion on Animal Planet that one time.

Helpful Advice

If you don't like the idea of your own skin being used to dismember your loved ones, try some of the following tips in the event of a hypercane.

1. Don't have a family.
2. Don't have skin.
3. Those are your only options.

It's the strength of the overall weather systems that matters; since that is increasing, that can mean some very, very bad things down the line.

A hypercane can form when a significant expanse of water reaches a temperature of about 120 degrees Fahrenheit, which is about 25 degrees higher than the highest ocean temperature ever recorded. But just because we haven't seen it doesn't mean it's impossible. Some theories hold that a hypercane was partially responsible for the extinction period that wiped out the dinosaurs. The meteor impact at Chicxulub in what is now the Gulf of Mexico would've started it all, but the resulting superheated ocean could have spawned a long-lasting hypercane that would have ravaged the Earth for weeks on end. The really scary part about the formation of hypercanes is that, much like the world's scariest bag of Lay's Potato Chips, you can't have just one: Any stretch of ocean superheated enough to spawn one hypercane would stay at that temperature long enough to spawn several more—so even though one is quite enough to kill the world just fine, it's brought all of its friends along . . . just in case.

Other Bad Things Down the Line

- Increased tidal activity
- More rogue waves
- Super-cell storm systems
- Twisters
- Complete disbanding of yacht clubs

A hypercane could vary anywhere from a measly ten miles in diameter to the size of an entire continent, but in the patronizing words of your ex-wife, "size doesn't matter, honey, it's how you use it," because even the smallest hypercane would have the same planet-killing effects as a continent-sized storm. A hypercane has winds of over five hundred miles an hour—more than enough to rip the skin from your body, not just to dismantle a house but to completely disintegrate it, and to send entire cities hurtling through the air like a normal hurricane would send trees. For a better idea of how devastating this event would be, think of it like this: A hypercane moves the very air around you at about the speed of a typical airliner. So the odds of surviving a hypercane would be about the same as surviving on a planet where the entire atmosphere—the very air itself—consisted solely of jumbo jets traveling at top speed. Clearly, your basement ain't gonna help much when you're trying to breathe in airplanes.

As an added bonus, a hypercane would also have a plume rising twenty miles right up into space. So if you ever dreamed of being an astronaut—now's your chance! You'll most likely be some form of jelly when you achieve that dream, but hey, we all make sacrifices for our goals, right? That plume is the truly worrying part: It would raise water, dirt, debris, and of course the obligatory trailer parks twenty miles straight up into the stratosphere. For those of you coming from public schools, that's like the bottom of space! This sudden influx of matter in the upper atmosphere would punch a hole right through the ozone layer and scatter everything formerly safe on the ground into orbit. On the plus side, suborbital trailer parks sound marginally more livable than normal trailer parks, but on the downside,

Tips to Survive a Hurricane

- Stay away from glass.
- Seek shelter in a basement or small room.
- Have an emergency kit prepared.

Tips to Survive a Hypercane

- Don't.

the debris would then act as a superpollutant, blocking out the sun, poisoning the air, and triggering even further planetary devastation. The water and dust molecules introduced to this fragile area would also block the atmosphere's ability to absorb harmful ultraviolet light. So hey, if you do manage to survive the actual hypercane with the power of clean living and intense prayer, you still get terminal space cancer if you ever see the sun again. Jesus, it's like it not only wants to kill you, but also plans to take away everything good about your life if it can't. The hypercane sounds so epically awful that it would have been equally at home in either science fiction or as a Care Bears villain—just out to steal joy away from the world.

And just when you thought it was over—well, it's quite possibly never going to be over. Because the extremely low barometric pressure inside a hypercane also gives it a nearly indefinite lifespan. For example, look skyward: See that giant spot on Jupiter, commonly called The Eye? The one that's been there for thousands of years? Technically, that's a hypercane. And if the conditions are exactly right, a self-sustaining infinite hypercane is also theoretically possible right here on Earth.

But hey, it's not so bad. After all, the hypercane takes a lot to be triggered: It needs a large expanse of water rapidly heated to well over 100 degrees to form. Anything capable of achieving something like that is pretty unlikely to occur. It would take another form of serious disaster, like a worldwide rise in temperature (a "Global Warming," if you will) or an asteroid impact like Apophis (see chapter 12) or an underwater supervolcano (see chapter 6) like on La Palma (see chapter 7) . . . to . . . shit.

Put on your screamin' shoes, looks like we're going hypercane shopping.

NANOTECH THREATS

Great leaps in human technological advancement are often initiated by the rise of a single, new, unforeseen field of invention. The forging of metals brought us solidly into the age of construction; the printing press brought us into the age of literacy; and the modern factory system brought us into the industrial revolution. The next big leap in technological advancement is, according to all sources, just over the horizon: nanotechnology. If industrialization made consumer products easier, cheaper, and more readily available, nanotech is going to make consumerism practically rain from the sky. Nanoparticles, the term for inert, nonmachine molecules reduced to the nanoscale, could theoretically do anything from eliminating cancer to creating self-mending clothes, while nanobots, the more complicated microscopic machines, could rearrange the building blocks of matter itself, essentially creating something out of nothing. It's going to be like having a million tiny robot butlers at your beck and call who live inside your body, and whose only desire in life is to fetch you as much awesome as you can hold.

This is the world of nanotech, and it'll be the best thing that ever happened to you . . . if it doesn't kill you first.

9.

GREEN GOO

COMPUTERS ARE reaching their saturation point in our everyday life—cell phones, iPods, digital cameras. They're getting more ubiquitous by the day and smaller by the minute. They provide most anything, from serious applications like military command and genome sequencing to the more trivial tasks like supplying online journals or easy access to obscure fetish porn. And it's no wonder they're so omnipresent; what other device could fill all those niches at once? What else could simultaneously function as an efficient soldier, run complex laboratory data, allow you to express your innermost feelings, *and* show you people fucking in cartoon coyote suits? Computers have thoroughly inundated modern life, so why not take it a step further and inundate your life, quite literally?

Well, nanobiotechnology—the term for nanotechnology applied to biological systems—proposes to do exactly that . . . and a lot sooner than you may think. You see, scientists are already implementing the first wave of human-altering nanomachines, and they expect to have the first legal, commercial applications available within the next decade. But the technology may be moving faster than we're able to fully understand it, and some issues that are already cropping up are, to put it politely, so terrifying that the fear shit you take will inexplicably shit *itself* in terror.

The "Green Goo" scenario is a theory stating that the true danger of nanotechnology lies not within the mites themselves, but in the creatures they modify. Much like the concerns surrounding genetically modified foods, the idea here is that any introduced trait that turns out to be beneficial will enter the gene flow and start to carry over naturally. It addresses such concerns as

This effect is similar to the theory of "trickle-down economics," except that instead of hoping that the superrich accidentally drizzle money over the poor like monetary salad dressing, in this case it's human-augmenting robots trickling into the ecosystem through waste by-products. Basically you're pooping superpowers into the swamp.

what might happen to an ecosystem if a strain of nanobots or nanoparticles accidentally improves, even marginally, something like the eyesight or immune system of a top predator. Furthermore, what could happen to the predator if its prey suddenly possesses, say, heightened endurance? These sorts of scenarios can't be fully tested in labs, because they deal expressly with nonlaboratory conditions taking place solely in wild ecosystems. And the effect is cumulative, so what may start with a harmless frog leaping just an inch higher could well end up with a sky eternally darkened by sinister patrols of helicopter sharks.

And as usual, it all starts very small, and with only the best of intentions: Researchers at the University of California have recently developed nanomites equipped with small doses of chemotherapy drugs. These simple nanobots actively target cancer that is attempting to spread, and then attach to a protein found on cancerous blood vessels that supplies tumors with their oxygen and other nutrients. They then inject their payload of drugs into the vessels, which causes them to deny the tumors sustenance, thus preventing the cancer from metastasizing and spreading to other organs, which is really what kills most cancer patients. The drugs don't eliminate the tumor; they just contain the cancer and starve it until somebody can come along and kill it. To put it more succinctly: They function like a million tiny Auschwitzes . . . inside your blood.

This development is important because the cancer drug used—doxorubicin—is also a highly toxic poison, one that causes fatal heart attacks in a significant portion of the people it's administered to. But the nanobots are able to administer the drug so precisely that the amount needed for treatment is drastically reduced, and the side effects are almost nonexistent. It's a technology that could save your life one day, and a damn good reason to have these things inside your blood and be quite happy about it. Just try to avoid thinking about the submolecular genocide raging inside your veins. And I wouldn't mention the fact that the poisonous robots living inside your blood are the only thing keeping you alive, if I were you. (There are nice, padded places the police tend to bring you to if you say things like that out loud.) Oh, and definitely don't dwell on the somewhat worrying prospect that if too many of these poisonous-drug-administering nanomachines were introduced into your body, just waiting for the cue to activate, you'd be basically walking around pre-murdered, just waiting for somebody to take a whack at the robot-filled poison piñata that is your

Historical Trade-offs That May Not Have Been Worth It

- Indians swapping Manhattan for beads
- The Chinese allowing British occupation in exchange for opium trade
- Curing cancer by injecting poisonous robots

body. So that Auschwitz-in-the-blood analogy from earlier still holds true. It's just that this time, you're the one in the showers.

But hey, don't start getting freaked out yet! This wasn't even the part intended to scare you; it was just an example of some of the wonderful things that nanotechnology can do for you (in this case, injectable anticancer Nazi robots).

It's very unlikely that the burgeoning field of nanobiotechnology will be reserved solely for medical uses. There's a pretty standard, preestablished pattern of dissemination in place for new technology. At first, new tech is *always* reserved for serious uses, but before long it's so commonplace that you have it everywhere. Take the internet, for example: It was exclusively a military network just over forty years ago, and now half the line at Starbucks is tapping into said former military network to check out grammatically impaired cats while waiting for their Grande Frappucino. So sure, nanobiotech is just for cancer treatment now, but maybe tomorrow it's a flu vaccine and maybe a week from now it's a pain reliever, or a subdermal sunblock, or maintenance-free contact lenses. But if it's that commonplace, then why not go further? Why not a customizable, morphing tattoo? Or a permanent cell phone in your ear? How about a remote control installed in your brain? The potential uses are tempting, and the appeal is easy to see.

We've established that it will be inside everybody eventually. Then what? Even supposing that nobody abuses the technology, its very nature makes even the most benevolent intent potentially lethal. Take the respirocytes proposed by nanoexpert Robert Freita: They're a harmless application, just a kind of artificial red blood cell that pumps oxygen more efficiently and more stably than the natural equivalent—236 times more efficiently, to be precise. Because of this dramatically heightened performance, they would be invaluable in treating disorders such as anemia and asthma, or simply to oxygenate the blood for better endurance and performance in sporting events. In other words, they're blood-borne nerd fixers.

However, with higher doses of these nanobots, their hosts could also be able to do things like hold their breath for several hours and run at a dead sprint for nearly twenty minutes. And that's great! How many of you would want to be able to do that? Now, how many of you want *anybody else* to be able to do that? Go ahead; raise your hand if you want sociopaths who breathe underwater, sprinting rapists, and serial killers who never tire.*

*Put your hands down, aspiring serial killers and rapists. Your vote does not count here.

Not to mention the worrying fact that these machines are by no means human-specific. The respirocytes take a kind of frat-boy approach to blood: If it's warm, it needs to be pumped. No further distinctions need to be made. Also of concern is the fact that their durable outer shell and self-sustaining programming make them seriously hardy devices, easily capable of surviving and functioning outside of their intended environment for long durations. And when you factor in how easily they could spread—their transferability by blood and other bodily fluids—you start to get a worrisome picture. One bad accident at the local zoo with somebody hosting these nanobots and next thing you know, you've got untiring, superspeed pythons racing through the streets and a terrifying new version of sea lion roaring at the bottom of your pool. In an instant, the food chain is drastically reordered. Though the chief concern for now is just the effect a modified species could have on its local ecosystem, any supercharge in the efficiency of predators is the last thing we need.

After all, humans are only at the top of our food chain because we're smart enough to compensate for our insane physical incompetence as a species. So . . . maybe you should start studying. Because pretty soon, a billion tiny robots might be seriously hot-rodding up some grizzly bears, and you? Well, let's just say you're going to have to get a hell of a lot smarter in a big hurry if you plan on making it back from the store with both arms.

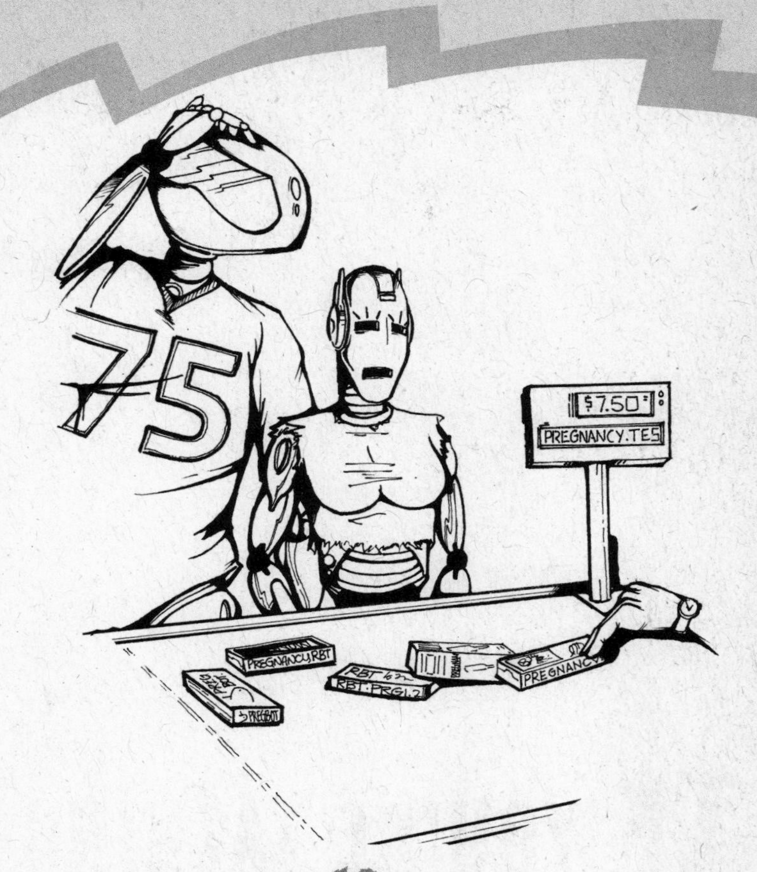

10.
GRAY GOO

IF YOU'RE TALKING about nanotechnology at a party, two things are assured:

1. Nobody is going to have sex with you in the foreseeable future, and . . .
2. Somebody will bring up the Gray Goo Scenario.

The term "Gray Goo," for those of you probably too busy boning right now to read this, describes the danger of self-replicating nanomachines running amok, forgoing any meaningful objectives in favor of just endlessly reproducing themselves like tiny little robotic Irish Catholics. The term was originally coined by a man named Eric Drexler in 1986, in his book *Engines of Creation*. He called it this *partly* because nanotech was just being recognized as the next industrial wave of the future, and also because the far more awesome title, "Engines of Destruction," was already taken by three Swedish metal bands, two monster trucks, and one particularly shitty mechanic.

In his book, Drexler writes of Gray Goo as something akin to the Midas touch, the simple wish that everything you touch turns to gold, which leads to you dying of starvation, because you cannot eat gold. Here the simple wish is that you didn't have to build every single goddamned microscopic robot by hand, which leads to your limbs being eaten by robots. Because if you encourage incredibly simple nanobots to build more of themselves, the danger is that they won't know when to stop pulling apart matter for its raw building materials, then using those appropriated materials to build more robots, which, in turn, will do the same thing. If left unchecked, the nanobots would eventually break down everything into its core elements, effectively restructuring the entire planet into robots. While the idea of an entire planet turning into a robot may indeed make a sweet-ass plot for the next Transformers movie, the

Hindsight Is 20/20

"I guess we shouldn't have 'encouraged' those robots to eat people in order to build more people-eating robots."
—Early nanotech scientist

unfortunate consequence would be the end of all life as we know it. Not exactly worth the trade-off, in my opinion.

The terrifying notion of microscopic organisms pulling apart base matter and assembling more dangerous creatures isn't exactly new. It was originally inspired by DNA, small molecules that break down raw materials and build more complex molecules from them. They gave structure to all life on Earth, and all a self-replicating nanobot does is follow this same concept, with the limiters pulled off. If you did the same thing to human beings, the results would be similar; we're basically just destructive machines tearing up shit to build more of ourselves until there is nothing left on Earth but a mass of writhing bodies engaged in a gargantuan, planetary-scale accidental orgy.

Luckily Eric Drexler wrote another essay years later that tells us that Gray Goo is just not ever going to happen. He assures us that there is simply no practical need for nanobots to be self-replicating,. because it would make far more sense to build tiny "nanofactories" that manufacture completely nonreplicating robots. The factories themselves wouldn't be autonomous; they'd be immobile and dependent on human resupply, so there would be absolutely no danger of infinite reproduction. Indeed, it would actually be *much harder* to engineer a single Gray Goo nanobot than it would these simple nanofactories. The factories themselves needn't be microscopic, after all; they're actually more likely to be somewhat large in scale—probably around the size of a photocopier. (You may want to keep this

Earnest Question

Then why did you invent the *entire concept,* Eric Drexler?

fact handy for future reference, lest ten years from now you find yourself at an office Christmas party, drunkenly attempting to photocopy your ass, but instead find that its base matter has been reappropriated into microscopic robots. The normal apocalypse is bad enough. We don't need one made entirely out of drunken cubicle jockeys' former butt cheeks.)

Potential Names for This Rear-End Apocalypse

- Asspocalypse
- Catasstrophe

So it's easier to build a factory to do the microscale work for you and, of course, there would also be less programming needed for the factory produced nanobots, because unlike the Gray Goo 'bots, they don't need to procreate, just work. So while that kind of sucks for the little robots (no robot sex for you little guys; there is only work until death), it is pretty good for us; no disintegrating, just magic invisible construction workers! I know that's a great concept, but do try to avoid saying that phrase out loud until this is common knowledge. It will only generate more odd stares at the person already laughing at a book about the apocalypse. You don't need the help.

Handy Tips to Stop a Nanotech Infestation in Your Body

- Live in a cave forever.
- Swing your arms about wildly, as if swatting invisible insects.
- Hide your blood.

But all these construction concerns pale in comparison to the fundamentals necessary for a Gray Goo 'bot to even function. Every nanobot would need to have five important capabilities, any one of which, if absent, would render the whole scenario impossible. First of all, to even acquire those building materials (read: your flesh) they would have to be mobile, and that's a heftier task than you might imagine. Second, power is an issue: How do you power the microscopic nanobot, much less its even smaller nanoengine? It's not like you pour thimbles of gas into its tiny fuel tank, as adorable as that would be; it needs an entirely new fuel source, and how do you come up with something like that that remains nontoxic to the human body?

Horrifyingly, that's how!

Scientists recently broke this fuel barrier when they started powering medical nanobots with the reengineered tails of human sperm. It's quite an elegant solution, really: Sperm are perfectly functioning, natural motors that power themselves solely on glucose, a chemical already naturally present in the human body. And this system also knocks down another barrier of the fundamental capabilities needed for a Goo-bot to function, which is combating the human metabolism. Glucose is a perfect fuel for any use within the human body, so these natural engines could be used on any biological robot, from cancer-fighting nanobots to system-enhancing modifications. Researchers are looking into using these things for anything from curing paralysis to reducing asthma. So on the plus side, there might just be a new weapon in the battle against disease. On the downside, that weapon is basically cum in your blood. Hey, at least homophobia will die off pretty quickly, when all bigots refuse treatment because their organs "ain't no queers."

Potential Upsides to a Gray Goo Infestation in Your Body

- You'll never be alone again.
- It probably tickles.
- Parts of you will live forever! (Or at least long enough to disassemble your loved ones.)

The third element a self-replicating nanobot would require to function would be an incredibly sturdy shell. Microscopic machines not only have to endure the intense atmospheric pressures of the body and atmosphere, but also need to fight off interference from sunlight, bacteria, temperature—basically everything. Sure, it's a hard-knock life, nanobots, but that's what you get for being a doomsday scenario. What did you expect, pity? You want to turn our stomachs into your children, nanobots; we have no sympathy for you here.

The fourth fundamental is control, because what's the point of having a roving, armored nanobot if it just wanders around aimlessly with no clear goal in mind, stumbling awkwardly through your blood with absolutely no purpose in life, like a tiny little robotic teenager?

And finally, the last fundamental is that of fabrication: A nanobot would need to constantly carry around all the tools it needs to build more of itself, in addition to the tools needed for whatever job it was engineered for. These things are smaller than bacteria, so there's not exactly room for a carry-on.

All of this adds up to a simple realization: There's just no call for a self-replicating nanobot. It is not only impractical, but actively dangerous. Why go through that if there's an easier, cheaper, safer solution?

See? Everything's cool.

No, really, buck up, friends! Those little robot fiends are impotent! Aside from tiny feelings of cybernetic frustration, a gaggle of wee unsatisfied female robots, and some itty-bitty inklings of machine inadequacy, there are no negative consequences!

Nobody dies today!

Why, Eric Drexler himself states that Gray Goo has become a scaremongering scenario that only takes away from more pressing concerns regarding nanotech. Chris Phoenix, Director of Research at the Center for Responsible Nanotechnology, also states quite clearly that it is a nonissue:

> Runaway replication would only be the product of a deliberate and difficult engineering process, not an accident. Far more serious, however, is the possibility that a large-scale and convenient manufacturing capacity could be used to make powerful nonreplicating weapons in unprecedented quantity, leading to an arms race or war. Policy investigation into the effects of molecular nanotechnology should consider deliberate abuse as a primary concern, and runaway replication as a more distant issue.

So that's . . . comforting, I guess? He's saying that Gray Goo won't ever happen on accident! Admittedly, it would be slightly more comforting if he didn't also say, practically in the same breath, that you shouldn't worry about Gray Goo happening *on accident*, because it's only going to happen *on purpose*, and even then only if *far more terrifying things* do not happen first. Jesus, hopefully nobody's turning to cry on Phoenix's shoulder, because he certainly didn't get his doctorate in compassion.

However, who would ever want to engineer it on purpose? Gray Goo wouldn't be appealing for military purposes because it's so hard to

Better Names for Gray Goo

- Steel Pudding
- Robot Death Buffet
- Reverse Voltron Disorder

control and so wantonly destructive. When other nanotech weapons can be used far more effectively to kill with control, who would want something that just randomly destroys life and sows chaos? Only psychopaths and terrorists want that kind of stuff.

Oh, wait; we have a whole bunch of those around, don't we?

In light of this fact, the Center for Responsible Nanotechnology realized that they couldn't scratch Gray Goo off their list of concerns just yet, but did add that it was a low-priority threat because there were "far more dangerous and imminent issues with nanotechnology."

There were far more dangerous issues than sperm-powered blood robots eating the Earth.

That's what they said.

They think that's comforting.

"I'm so sorry, sir. You have terminal cancer. But don't worry; it's all going to be OK! You won't be dying from the cancer . . . because I am going to shoot you in the face right now. Isn't that reassuring?"

No, Center for Responsible Nanotechnology, that is not comforting. I would suggest you offer hugs instead of these horrifying press releases, but judging by your previous "consolation" track record, I'm afraid you'd just end up whispering obscenities in people's ears while punching their children.

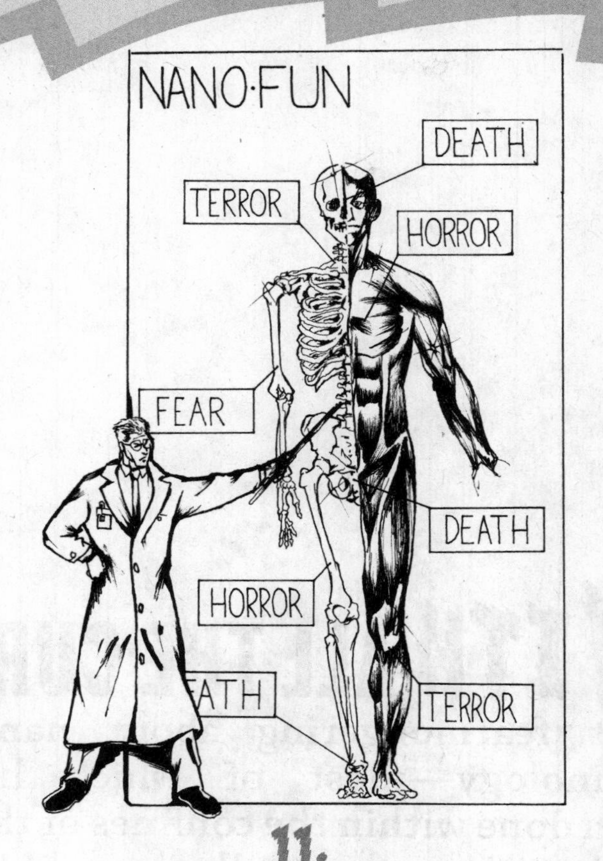

11. NANOLITTER

WITH ALL THE CUR-rent fearmongering about nanotechnology—most of which has been done within the confines of this book—it's not *actually* very likely that the nanobots are going to build their children out of the last sad remnants of Earth or inspire a deadly new team of superanimals like the world's tiniest Legion of Doom.

But don't worry, that doesn't mean we're not all going to die anyway!

This is because even the most benevolent of nanobots share one simple, undeniable commonality: To achieve any major effects, there are going to have to be a lot of them and, though it is infinitesimal, they do take up some space. When their purpose is complete, they'll deactivate and die off—unfortunately leaving their corpses where they lie.

But so what? Aside from the hassle of having to construct a plethora of mini tombstones (and the rather gross prospect of microscopic widows getting grief fucked inside your teeth), how could this possibly affect you? Well, the primary application for a lot of this nanotech is going to be for health issues: improving stamina, boosting immune systems, and fighting off cancer. All those corpses are basically just litter, and the human body is their environment, which means that when they die, they die *inside of you*. And if you think this scenario—that you'll be poisoned to death by the corpses of miniature robots that live inside your blood just like that hobo down by the library keeps screaming—sounds somewhat outlandish, then you should know one little thing: It's already happening.

The Five Stages of Grief

- Denial
- Anger
- Bargaining
- Acceptance
- Grief fucking

Take, for instance, the scientists over at Kraft foods—scientists whose job, ordinarily, probably consists of formulating the perfect Dora the Explorer pasta shape–to-cheese ratio—are instead currently working on new types of nanoparticles to add to beverages. They plan to create "interactive beverages" that will shift colors and patterns according to your input. So on the plus side, you can have green beer whenever you want it, but the trade-off is that it's *potentially full of superpoisons*. Some would argue that these risks dramatically outweigh the benefits here, but those people probably haven't spent their entire lives wishing beyond hope that their Coke would turn pink when they rubbed it. Clearly, those bastards just don't understand the dream.

Kraft Interactive Mood Beverages

- Angry Cherry
- Depression Blueberry
- Jealous-Rage Sour Apple
- On-the-Prowl Pink
- Willing-to-Settle Gray

OK, so nanoparticles aren't exactly murderous microscopic robots. They actually have a lot of positive effects, and are being used in vastly increasing numbers across myriad products, from paint to socks, makeup to underwear. Their upsides are easy to see: They can have a lot of useful effects, cost very little to produce, and take up almost no space in existing products. Most of what we know about them is quite beneficial; it's the stuff we don't know that's worrisome.

Back in March 2002, the EPA found the first inklings of the problems to come, when a study they'd conducted found nanoparticles cropping up in the livers of research animals. This warranted urgent further study, seeing as how nanotech was on the verge of becoming the world's largest emerging industry. Researchers at the University of Rochester Medical Center quickly confirmed that at least one kind of nanoparticle could indeed penetrate the skin, and from there seep into the bloodstream. Those particles are called quantum dots, and they're on the smaller end of the nanoparticle scale. They are often used in makeup and sunblock, which is unfortunate, considering how they seep through skin like that—but even more unfortunate when you consider that UV light, like from the sun, actually facilitates absorption of the dots. So the thing you use to protect yourself from the sun is actually rendered harmful and then activated and inserted into your body by the mere presence of sunlight. Apparently the engineers in charge of quantum dot production got their doctorates in Irony from Incompetent University.

Separate research conducted by scientists at Purdue University concentrated on tracking the likelihood of other nanoparticles, called buckyballs, infiltrating human systems—be it through water, soil, or the fatty tissue of the livestock we consume. And they found that there was indeed a pretty high chance of these buckyballs attaching to our own fatty tissues—even more so than DDT, the notoriously harmful pesticide. Now, to be

Rejected Slogan for Nanoparticles

"Pentetratin' you like you know you like it since 2002."

fair, it was not outright stated in the study that buckyballs do anything worse than DDT once they get in there, but this comparison *was* specified in the report. That's like conducting research that concludes that adorable bunnies are ten times more likely to be found in your home than murderous serial killers. Sure, it's innocuous enough information, but when you phrase it like that it's clearly going to scare the shit out of everybody.

But nobody was entirely sure what these revelations really meant for humanity at the time; as one panel member, Vicki Colvin, professor and director of the Center for Biological and Environmental Nanotechnology at Rice University in Texas, put it:

> One thing we've concluded is whatever these things [nanomaterials] are going to do, they're not inert. What will they do when they get in the environment, and what will they do when they get into people?

Other Disturbing Ways to Phrase Test Results

- "You've got more white blood cells than a vampire Klan meeting."
- "You've got a higher semen count than your whorish mother's mouth."
- "You've got an 'A'! And an 'I' and a 'D' and an 'S'!"

If seeing that sort of fearful uncertainty from people who typically really know their shit has you a little worried—don't be! That worry is totally premature; I would save it for later . . . when things get worse.

When the EPA finally decided that this stuff needed to be regulated way back in 2008, they started the Nanoscale Materials Stewardship Program, which requested that companies send in safety records for their environmental research efforts in the field of nanotechnology. The only catch? This program was completely voluntary, and the companies could omit literally anything they didn't feel like sharing. So basically the EPA asked large, profit-motivated companies to pinky swear that "everything was cool," and then followed up by asking them if it was "for realsies." And if there's one thing that massive corporations have shown they take seriously, it's the honor system.

Later, a member of the EPA council, Mark Wiesner, director of the Center for the Environmental Implications of Nanotechnology at Duke University and former director of the Environmental and Energy Systems Institute at Rice University, took the art of issuing worryingly ambiguous statements even further when he went on record with his concerns about large-scale nanoproduction, stating, "People talk about incorporating nanotubes in composites that might be used in tires. When you drive tires around, they wear down, and so nanotubes will be passed around in the environment. Where does this stuff go? What will be its interaction with the environment? Is it the next best thing to sliced bread, or the next asbestos?"

The Scale of Greatness to Tragedy, According to the EPA Council

- Sliced bread
- Kittens
- Warm cookies
- Alan Thicke
- Mayonnaise
- Awkward high fives
- Asbestos

So he's really just trying to say that there might be cause for concern here; he just worded it as vaguely, and as threateningly, as possible. He might as well state that this scenario is either a chocolate bar or a hand grenade; it's either a new puppy or a furious grizzly bear; either multiple orgasms or blindfolded chain-saw surgery.

As far from comforting as Wiesner's creepy PR statements are, it starts to get worse when you realize that he's being literal. See, carbon nanotubes closely resemble asbestos fibers in shape: They're elongated, thin, and bar shaped. But the tubes are not typically as dangerous as asbestos, because they have a tendency to group together, which alters their overall shape and thus renders them harmless. However, if they do split into single fibers, they can then inflict the same kind of damage that long-term exposure to asbestos has, like serious respiratory problems, and even cancer, say the results of a 2008 study published in the journal *Nature Nanotechnology*. So it's probably not such a great thing that a large area of current nanotech research is dedicated solely to finding methods that make sure these things *don't* clump together, but rather stay separated into their tiny, thin, deadly form. After all, asbestos wasn't all bad: Before it started killing people, it made somebody like a billion dollars!

Concerns are also being raised about interaction between nanoscale products and external factors. Even

Transcript from Development Meeting of Nanotube Shape Deciders

"Listen, guys, I love this tiny shit we're doing here, but I've got an idea and I'm just gonna throw it out there: Let's make 'em long, straight, thin filaments — like asbestos!"

"But sir, is it really the best idea to model our product after a deadly carcinogen?"

"You know what else is a deadly carcinogen? Your butt! Ho! Served!"

"Well played, sir. Cancer tubes it is."

if the nanotech itself is totally safe, if it meets up with the wrong stuff inside your body, all hell breaks loose. It's like this: Say you have a lovely, pure, angelic daughter. She's very talented, well-spoken, and a pleasure to be around. She's the light of your life. But one day she brings home her new boyfriend to meet the folks . . . and he's a rabid grizzly bear. Now, your daughter is still an angel in her natural environment (your home) and a grizzly bear is a noble, majestic creature in its home (the wild), but when you bring the two together, it tends to fuck up the family reunion. Now replace the family reunion in that metaphor with your own sweet internal organs, and you've got a delicious terror sauce that pairs well with both anxiety *and* horror!

Take gallium arsenide, for example: It's just a harmless semiconductor, kind of like a faster version of silicon. It's already all over many small-scale electronics and solar panels, but if you deploy it at the nano scale, suddenly it starts seeping into your body. And that's when shit really goes wrong, because gallium arsenide is made from gallium . . . and arsenic! You know, like the deadly poison? So yes, gallium arsenide is a completely safe tool on the normal scale, but if it starts actually getting inside your body, you might have some problems, and deadly, poisonous problems are among the shittiest genre of problems to have inflicted on every cell in your body. That risk is not inherent to just gallium arsenide,

Terror Sauce Wine Pairing Guide

- Red meat + terror sauce = pinot noir
- Fish + terror sauce = chardonnay
- Spiders + terror sauce = tears
- Clowns + terror sauce = nightmares
- Spider clowns + terror sauce = nocturnal fear emissions

either; pretty much any material could, theoretically, have that ill effect when brought to the micro scale. A toaster, for example, is a lovely machine—who doesn't love its warm, crispy ejaculations? But if you shrink that toaster down to the nano scale, well, suddenly it's a different story when that comforting, crusty breakfast staple is being made a billion times a second *inside your own heart*.

And it's not like you can just opt out of it, either, because this seepage factor means it won't affect only the voluntary users of nanotech. A core principle of effective nanotechnology, after all, is the ability to spread in anything from bodily fluids to simple skin contact, through our food supply, or even as airborne contaminants. And once they do enter your bloodstream, any number of other disastrous interactions can occur: The proteins in your blood may "wrap them up," thus distorting their own shape in the process. And when their shape changes, so does their function. Depending on what shape they're surrounding, those proteins may suddenly switch functions. They may, for instance, get confused and switch to clotting—causing your blood to suddenly coagulate inside your veins.

Or even more worryingly, bacteria may piggyback on nanoparticles intended for medicinal purposes. This is particularly bad because medicinal nanotech will be engineered to bypass your immune system, seeing as how your immune system would destroy the particles as it would any other foreign invader, and they wouldn't be effective as medicines if they were destroyed. So any bacteria piggybacking on these beneficial particles could then use said particles like tiny little BattleMechs—their otherwise weak bodies being shielded by the hardy, nigh-indestructible armor that is the medicinal nanoparticle. This would transform otherwise easily destroyed bacteria into little blood-borne ninjas, free to wreak severe devastation on your immune system with no way of being detected.

For a current example of potentially dangerous nanotech particles in use, consider nanosilver: It's used for its antimicrobial properties to both eliminate odor and reduce the chance of infection. As such, they're being mass produced for use in socks, underwear, bandages, cookware—a billion little particles with a billion potential uses, and they don't even need to be modified, just shrunk down. This is because some elements, when reduced to nanoscale, can suddenly have effects previously unseen in their large-scale counterparts. One gram of any nanoparticle less than ten nanometers in diameter is roughly one hundred times more reactive than a gram of the same material comprised of larger, micrometer particles. In short, the more you shrink it, the more crazy shit it does. Just like the Japanese.

It's also astoundingly hard to measure the exact effects of any random nanoparticles you may absorb, because the more you accumulate of any given particle, the more you change the way it affects you. In the case of regular-scale silver, the side effects on human beings are relatively harmless. At most, if you consume too much normal silver you'll develop argyria—a condition that turns your skin blue. It is permanent, but otherwise relatively harmless. The upside of normal-scale medicinal silver? It's an effective antimicrobial ingredient

Unnecessary Joke Explanation

Because the Japanese are typically a shorter people, and are, as a nation, batshit insane. God love them for their awesome robots, but you cannot dispute the epicness of their crazy.

whose flexibility and relative safety have proven incredibly useful to human beings. The downside? You might have to spend the rest of your life as a Smurf. And that's pretty OK, right? They seem happy enough folk, even if it is a bit of a sausage party.

Nanoscale silver is still beneficial, of course: It still has all the antimicrobial properties of its larger counterpart . . . it's just that it might have *too* much. Professor Zhiqiang Hu, from the University of Missouri, has conducted studies that showed that even relatively small doses of nanosilver can kill the bacteria used to process sewage and waste. So the more nanosilver you flush, the more invincible you make your own poop—a disturbing thought if ever there was one. And these aren't the only necessary bacteria that silver puts at risk: If too much gets into ordinary soil, it can eliminate nitrogen-fixing bacteria in there as well. All plants on Earth need that stuff to live, so if you kill that off, no more food for you. And as a human being, *you* probably need *that* stuff to live. Well, unless you've drunk too much normal-scale silver, in which case you'd be just fine; Smurfberries will be largely unaffected.

Downsides of Being a Smurf

Physically frail
Whole life limited by adjective before name
Always getting captured by asshat Gargamel
Only one woman
Sloppy 242nds

Yet another problem lies in the very nature of nanotech's construction. See, nanobots have to be made of only the hardiest materials in order to withstand the vast atmospheric pressures that would otherwise crush their delicate machinery. Materials like diamond, carbon, and even gold are used in pretty much all nanotech. Durable materials. Strong materials. Materials that do not break down. Materials that sit inside your veins, and just build up, and up. It takes only a millimeter of arterial plaque in your veins to provoke coronary artery disease (the leading cause of fatal heart attacks), and though nanobots are much smaller than that, there's going to be a hell of a lot of them. Basically, you could now be looking at a massively contagious worldwide heart attack. It's a supreme twist of irony: By developing microscopic, disposable machines in order to do away with the arcane, polluting, industrial practices of yesteryear, we may literally pollute ourselves to death from the inside out with the litter of the future. On the plus side, though, that litter is mostly made of diamonds and gold—so at least your insides will be blinged out like Snoop Dogg's car on the submolecular level. It's like they say: "Live fast, die young, leave behind a beautiful, jewel-encrusted cardiovascular system."

SPACE DISASTERS

Asteroids, radiation, frigid vacuums, and hostile aliens—let's face it: space sucks, sometimes literally. Space doesn't bring you flowers, or nurture abandoned puppies back to health. Space doesn't provide delicious sandwiches at the company picnic or help old ladies across the street. It doesn't do one damn nice thing for you; it basically just plots your death from the abyssal void of nothingness. Sinister threats from outer space may seem like science fiction to you, but it's only science fiction until it's *landing on your damn head*. Also, if you really stop and think about it, there's a lot more of space than there are of us.

My God . . . don't . . . don't look now, but I think it's *everywhere*. Space has got us surrounded!

12.

ASTEROIDS AND EXTINCTION-LEVEL EVENTS

AN EXTINCTION-LEVEL Event (ELE) is a massive die-off of the majority of life on our planet, and they often seem to be caused by a particularly devastating asteroid impact. It's not exactly a subtle or mysterious phenomenon. In a nutshell: big rock, big explosion. There's not much to do but die as hard as you possibly can. When most people think of major meteor strikes, they typically think of distant prehistoric events, like the one that caused the extinction of the dinosaurs 65 million years ago when an asteroid roughly six miles in diameter struck the Earth at a place called Chicxulub, which we now call Mexico, and began the most dramatic extinction in history. (It was not the largest extinction period: That dubious honor falls to the Permian-Triassic extinction event. But while the P-Tr event killed off most of the world's insects, the Chicxulub event managed to slay every single real live dragon at once, and that's the kind of dramatic flair that squashing a trillion bugs just does not possess.) Because we associate ELEs with such disasters in the long-distant past, the tendency is to think that catastrophic asteroid strikes are strictly relegated to ancient history when, in reality, nothing could be further from the truth. Meteors hit the Earth like your dad hits the bottle every time you disappoint him, which is to say very often, and very, very hard.

For example, see March 22, 2008, when a one-thousand-foot diameter asteroid passed within four hundred thousand miles of Earth—missing us by only six hours. To us, numbers like four hundred thousand seem vast, but in terms of space travel that's basically like being in Earth's pocket, and while missing something by six hours may seem like a lot to you, in astronomical terms that's practically already inside of you: easing just the tip of its disaster member in to see how you like it before the full-fledged catastrophic shafting begins.

But even if it hadn't missed us, Earth's atmosphere typically protects us from a good deal of the debris that space is constantly trying to murder everybody with, and when a meteor enters the atmosphere it usually results in little more than a pleasant shooting star. Wishes are made, boys become real, and everybody learns a little lesson about love, right? Well, those dramatic shooting stars typically come from objects no bigger than a grain of sand, and if a grain of sand can light up the night sky—while simultaneously giving life to the hopes and dreams of optimistic children throughout the world—you can probably imagine what might happen when something a thousand feet across comes barreling through the atmosphere. (Hint: It ain't granting wishes. Unless you're wishing for a painful and fiery death.)

The Best Wish to Make upon a Falling Star

"I wish that was not a meteor about to kill everybody I love."

If that asteroid does enter Earth's atmosphere, a variety of things can go down, depending on its specific construction. The heavier bodies, like iron-laden rocks, are the ones most likely to actually impact the planet. That impact would throw up insane amounts of debris, release levels of destruction akin to several nuclear bombs, and leave a permanent terrain-changing impact crater for thousands of years. The more loosely constructed dust and ice asteroids, however, can't always take the increased pressure from Earth's atmosphere, and usually explode *before* impacting. That kind of sounds like the preferred scenario between the two: If it doesn't hit, that's like we're getting off light, right? Not really. An object detonating in the air can actually do quite a bit more damage than a physical impact. The asteroid that missed us by a blink of an eye, for example, was a loosely constructed object; if it had entered our atmosphere, it would have detonated with a strength estimated at seven to eight hundred megatons. That's about fifteen times the strength of the largest nuclear blast ever recorded! With that in mind, it's probably safe to say that if a medium- to large-sized asteroid ever does make it through the atmosphere, we're all pretty well fucked, because our best-case scenario in that situation is for the meteor to hit us so hard that it changes the very Earth itself. It gets a little hard to be optimistic when that kind of destruction is the most you can hope for. But if you think you *can* still see a bright side in all of this, be careful; it could just be a blinding flash from the largest explosion in history.

Things You'll Have Time to Say Between Noticing an Incoming Meteor and Death

- "I alw— "
- "Get u— "
- "Oh go— "
- "Whatthefuckisthat?"

Unfortunately, if an asteroid is on a direct collision course with Earth, that very fact makes it less likely we'll be able to see it until it's far too late. Typically, we track asteroids by virtue of their movement parallel to us. But when they're coming right for us, we can't see them moving. They just look like beautiful, harmless specks of light. But even more worrisome are the asteroids already inside Earth's orbital path—the ones whose path the Earth is intersecting with regularly, the ones closest to us, the ones most likely to hit; we can't see those asteroids because they're so close to us that they're backlit by the sun. Remember the old campfire horror story about the babysitter trying to trace the threatening phone calls she's been receiving? Well, that babysitter is us, and that serial killer is the asteroids, and good lord! I—I hate to break it to you, but . . . those phone calls are coming from *inside the house*.

To give us a better shot at avoiding secret, invisible, flaming space death, a team of researchers in Canada is launching a small satellite telescope to help us spot these near-orbit asteroids better, but it's a low-budget venture and it could do only so much. And while something is always better than nothing, keep in mind that there are more than five thousand asteroids dangerously close to Earth that have already been discovered using just our meager existing technology—it's kind of hard to get stoked about the mere *possibility* of the *potential* to *maybe* spot a *few* more. Launching a satellite to slightly extend our sight range is like wearing a bulletproof vest when more than five thousand guns are pointing at you—yes, a few of those bullets are gonna be stopped, but any way you cut it, somebody's winding up as meat Jell-O in this situation, and we're a planet-sized target.

But hey, don't worry, the government is totally on this one: A more official (well, more official than Canada anyway) approach is already under way. The U.S. Congress has introduced the NEO Preparedness Act, a bill mandating that we create a special program called the Office of Potentially Hazardous Near-Earth Object Preparedness, which would develop the technology to track 90 percent of all near-earth objects (NEOs), even those as small as 140 meters, by the year 2020. You better believe NASA's on that shit too; they've decided that we would need a much larger version of Canada's tracking satellite in place, preferably near Venus's orbit, to achieve this Congress-mandated goal. Unfortunately, it would cost about 1.1 billion dollars for fifteen years of operation, and that's just not in NASA's budget. Also unfortunately, Congress is far too busy asking if baseball players are really as strong as they seem and trying to choke bankers with wads of cash to grant more funds to such trifling matters as the avoidance of space bullets, so they won't give NASA the money. NASA scientists have stated that they intend to get to work on pursuing other, less costly plans, but seeing as how Congress is probably scheduling appointments to review whether wrestling is real and appointing a committee to decide exactly how awesome the last season of *LOST* is going to be, NASA probably shouldn't hold their breath on this whole "averting Armageddon" thing.

But maybe we're getting ahead of ourselves. The last one of these asteroids to initiate an extinction-level event was more than 65 million ago, so how present is this danger, really?

Well, in 1908 we had a little practice drill for an ELE when a chunk of rock half the size of a football stadium exploded over Siberia, initiating a blast with the strength of about fifteen megatons—roughly one thousand times the strength of the bomb that fell on Hiroshima. With temperatures reaching 5,000 degrees, the impact destroyed two thousand square kilometers of forest—literally laying the trees out flat on their sides in enormous radial circles like a satanic Spirograph.

One witness, stationed at a local trading post, described what he saw:

> Suddenly in the north sky . . . the sky was split in two, and high above the forest the whole northern part of the sky appeared covered with fire. . . . At that moment there was a bang in the sky and a mighty crash. . . . The crash was followed by a noise like stones falling from the sky, or of guns firing. The earth trembled.

This man witnessed an event *so traumatic* that he could only speak in biblical terms afterward.

It was ominously phrased prophecies of doom delivered

by traumatized, grizzled old Russians like that that spurred interest in the NEO Preparedness Office, which would not only track future potential meteor impacts through orbital telescopes, but is intended to help research and fund a plethora of solutions. Apparently operating under the Kitchen Sink philosophy of panicking and throwing everything we have at any potential threat, NASA proposes to use everything from "gravity tractors" to "a shit-ton of nuclear missiles" to deter impacts. That gravity tractor idea sounds pretty crazy, but really it's just a plan to send a spaceship to tow the asteroid away. Some other proposed solutions, like firing a solar laser at it, or wrapping it entirely in plastic like a planetary Hot Pocket, are far more bizarre. The most practical solution on the table is that aforementioned nuclear blast, but there's a major problem that prevents us from even nuking the damn thing: Blasting an asteroid apart preimpact could just fragment it into thousands of smaller but still Earth-impacting meteoroids. So now instead of a punch to the face, we've turned it into a shotgun blast. A nuclear shotgun blast aimed right at us. And that's our best option!

But one possible solution to this problem is the concept of nuclear pulse propulsion—essentially, using nuclear blasts as a kind of engine to push the meteor away from us without damaging it. They want to use

Brainstorming Notes on How to Repel Incoming Meteor Strikes from NASA NEO Meeting

- Blow the fucking thing up.
- Divert it.
- Like, push the Earth out of the way somehow?
- Last idea is stupid. How can you push a planet?
- Bigger gravity tractor.
- Invent Hulk, have Hulk punch.
- Ask Jesus.

nuclear explosions as the *fuel* in a gargantuan space engine that they'll attach to an incoming planet-killing asteroid. So really, if you can gauge the level of a threat based on the truly, epically insane lengths people are willing to go to prevent it, then you should probably start burrowing underground and asking every woman you see if she'd like to repopulate the Earth with you, because the only thing crazier than the plot of a Bruce Willis movie coming to life and ending the planet is the psychotic ways the government is trying to stop it.

And even if nature, fate, and God don't conspire to seal our fates with a giant rock kiss, we just might do it ourselves. Carl Sagan, in his book *Pale Blue Dot*, reasoned that any method capable of turning meteors away from Earth could ultimately be just as effective at rerouting otherwise harmless asteroids toward us. Sagan thought that since political leaders are all basically batshit insane, Earth will be at greater risk from a man-made impact than from anything naturally occurring. So he believed that by introducing ideas meant to avert disaster, we would actually give the bad guys some ideas to invite that same disaster. As if to prove his point, the Soviet Union read his theories and immediately set about work on Project: Ivan's Hammer, a military operation whose sole purpose was the complete weaponization of space by steering incoming asteroids toward specific global targets. Sagan was immediately struck dead by the irony. May he rest in peace, though he's far more likely spinning in his grave.

So even if the randomness of space doesn't kill us, there are people on Earth more than willing to take up the slack? I guess it's really only a matter of time before it happens. It could be any minute . . . it could happen . . . right . . . *now*! Ah, just kidding.

You've got until 2029.

Oh, I'm sorry, what? You didn't know? That's when the next one might hit. It's named Apophis, and it's very excited to meet you. It would shake your hand, but it prefers to say hello more the old-fashioned way: with explosions. But after that, shall we say, *warm welcome*, the conversation might turn *chilly* when the impact winter sets in! That is, assuming you don't die from *awesome pun overload* first . . . ahem. So anyway, Apophis is expected to pass dangerously close to the Earth in early 2029—closer even than our own geosynchronous satellites! And though leading scientists say it's unlikely to hit based on their projections, with a probability of only about 1 in 45,000, they also mention that their projections at this point are "not an exact science," which, when you think about it, is a pretty shitty thing to hear from an astrophysicist. Add the fact that these trajectories are easily influenced by any and all outside force—from planetary pull to space junk (you know, like those geosynchronous satellites it will be passing straight through)—and it's still somewhat unlikely that Apophis will hit in 2029. But any alteration of its course resulting from those satellite impacts could result in it hitting the Earth the next time it comes around . . . in 2036.

Apophis isn't going to be an asteroid one-night stand, scaring you just the one time and disappearing forever. No, this is more like an as-

Events with a Probability of About 1 in 45,000

- Stubbing both toes in the same day on the same thing.
- Finding a $20 bill on the street.
- Winning fifth prize on Scratch-it.
- All life on Earth being blown to holy shit by an asteroid twenty years from now.

teroid *relationship*, and my friends, I'm sorry to say that it is a dysfunctional one. If you're the kind of person who likes to look on the upside, though, you could think of it this way: It's like a bonus! You looked in the box expecting only one, but now you've got two free, heaping scoops of explosive death in every box of your terror flakes. There's also a supersecret prize inside. (Hint: It's more explosions.)

13.
VERNESHOT

THE EARTH IS A gun, and your country is a bullet. No, those aren't poorly translated Japanese metal band lyrics, nor are they the pseudo-poetic mewling of jilted emo children; those words could be, terrifyingly enough, a completely true statement. It's all because of something called a Verneshot, and though the theory is still under debate, it is the only one so far that explains why mass extinctions, severe geological damage, and volcanic eruptions often occur simultaneously all throughout history. Not content to simply state that "some shit went down," scientists have instead begun pointing to the Verneshot. And then probably screaming. And then dying.

That's just what the Verneshot does.

The big extinction that we all know about—a meteor killed the dinosaurs—is referred to as the Cretaceous-Tertiary, or K-T, extinction. New theories suggest that that disaster could have been caused by a Verneshot rather than a meteor impact. A team of scientists led by Jason Phipps Morgan at the GEOMAR Earth Science Institute at Kiel University first proposed the theory, which goes like this: Huge volumes of volcanic gas slowly build up beneath layers of impenetrable rock, called cratons. When those rocks start splitting apart ever so slightly, the built-up gases explode through the weak points—blowing the craton into a suborbital trajectory. The expelled chunk of rock is launched into the air, orbits the Earth briefly, and then crashes back to the planet with nearly the force of a meteor impact. Meanwhile, the tube that formerly held all the gas has emptied, pouring its noxious contents into the atmosphere. It then collapses in upon itself, causing an earthquake.

It sounds like just a more extreme version of a volcanic eruption—big

Fart Jokes That I Could Have Made

- "So much gas is released, it's like the Earth ate a seven-layer burrito."
- "More gas is released into the atmosphere than your grandpa's La-Z-Boy."
- "So much gas is released, it's like somebody punctured Michael Moore."

rock, gases, seismic activity—but the twist is in the scale of the thing: See, cratons are usually gargantuan. About the size of a country, to be exact, and that's a bad size for something that's being shot at your face. But the rock isn't your only worry: The tube that launched it—also hundreds of miles wide—causes devastating earthquakes upon its collapse. Estimates show that these earthquakes are off the current charts, estimated at an 11 on the Richter scale; the scientists in charge of measuring this would have to create a new notch on the dial just to do so, making them the Spiñal Tap of the earth sciences. So much gas is released that it poisons the entire atmosphere for thousands of years, blotting out the sun and corrupting the air itself.

But hey, let's not get distracted here; there's still a small continent in the sky that wants you dead. Let's get back to that, shall we? Upon impact, the blast would be akin to 7 *million* atomic bombs going off in the same place, and at the same exact time. That's too big a number for too bad a thing for most of us to fully comprehend. So if it helps, picture this: The city of London has a population of roughly 7 million. So imagine that the entire city of London is populated by atomic bombs. Atomic bombs in place of secretaries, gas station attendants, and schoolchildren—everybody. Exactly the same as London in every respect, but instead of each individual person living there, there is a device with exactly enough power to destroy Hiroshima. And then one of them trips.

How to Survive a Verneshot in One Easy Step

1. Don't live on a continent.

Charmingly enough, the Verneshot was named for Jules

Verne, whose book *From the Earth to the Moon* posited that space travel could be accomplished by loading astronauts into a giant cannon and just firing them at the lunar surface, presumably operating under the theory that the moon is made out of pillows. And it is, after all, a pretty fitting name. Because in light of all we've learned, it is technically possible for a huge cannon to shoot you into space; it's just that you'll be screaming particles of horror and guts when you get there. That probably would have made a much shorter book, though.

Some scientists dramatically explain the Verneshot as being akin to the Earth "shooting itself in the head." But perhaps that analogy could be more accurate: It's more like the Earth chopping off its own hand and then punching itself to death with it. Because bullets are for pussies.

But the scientists aren't just pointing at the possibility of a Verneshot because it would make an excellent premise for a Michael Bay movie. They actually have this thing they call "evidence." The Kiel University scientists say that the K-T isn't the only massive die-off possibly caused by a Verneshot. There have actually been four major mass extinctions that coincide with potential Verneshot scenarios since life first appeared on

An Excerpt from the Realistic Version of *From the Earth to the Moon*

Charles straightened his protective impact-derby, mounted the atmos-cannon, and bid a formal farewell laced with restrained emotion to his most loyal and loving children. There was a sound like the bellow of Gabriel's trumpet upon God's return, and my dear beloved Charles ventured forth into the cosmos. As something akin to hamburger. In retrospect, this idea was poorly thought out, at best.

Earth: the Late Devonian extinction event about 364 million years ago, the Permian-Triassic event 251 million years ago, the Triassic-Jurassic event 200 million years ago, and the aforementioned K-T event 65 million years ago.

The bizarre thing that first tipped off the Verneshot scientists is that these extinctions all have something in common: Available evidence seems to indicate that they were not only preceded by a massive meteor strike, but that there was always a simultaneous appearance of continental flood basalt, which coats great swathes of the Earth in liquefied basalt lava, forming dramatic landscapes and releasing massive quantities of poisonous gas in the process. I shouldn't need to tell you that the odds of two mass-extinction-causing events are extremely low (about one in 3,500), but it looks like I just did, doesn't it?

A lone extinction linked to the one-two punch of a meteor strike and a flood basalt flow? That's unlucky, sure, but shit happens. However, four instances of species-destroying simultaneous disasters? Well, clearly a new theory is required to explain when two such large-scale disasters seem to be occurring in concert. Because the only other sensible explanation, that global disasters like to gangbang the Earth like an aging porn star desperate for rent money, just suggests a God too perverted and cruel for the human mind to comprehend. So, rather than believe that God uses double-header disasters to fuck life out of existence, there are some scientists who would like to politely suggest that the Verneshot, not a meteor strike and flood basalt flow, is the more reasonable explanation, if only to retain one's sanity.

Of course, it is just a theory. Most ideas in science technically *have* to be labeled "theory"—which you can see in everything from relativity to evolution. Absolute proof is such a tricky thing to come by in the best of cases, much less when you're trying to prove the existence of something that not only would have exploded most of its evidence, but shot half of it into space afterward, then buried whatever scraps were left under continent-blanketing lava. Regardless, some hard evidence is being turned up that helps validate the Verneshot theory: Beneath almost all of the continental flood basalt, scientists are finding concentric circles engraved in the earth on a scale so large that it defies comprehension. Enormous furrows that get both deeper and narrower the closer they lie to the center, creating an inverted cone leading to one central point in the sea floor. A focal point of impact surrounded by the debris of a gargantuan collapsed tube. Sounds a lot like the exploded barrels of Earth-shatteringly huge cannons . . .

And though the most definitive proof of a Verneshot would be, much like in normal forensic investigation, to find the actual bullet, I should remind you that oftentimes it's enough simply to produce the gun in court. Considering that people are currently living on the murder weapon, it shouldn't be too hard to find. And so the Verneshot theory is gaining ground in the scientific community—ground that it probably just loads up and fires

QUICK QUIZ

Verneshot Facts, or Things Screamed by Crazy Hobos at the Bus Staton?

1. "India murdered Mexico."
2. "Where my dick at?!"
3. "Japan is a concealed weapon."
4. "They take your thoughts out like recycling!"
5. "The United States is ammunition."

Answer key: 1—Verneshot, 2—hobo, 3—Verneshot, 4—hobo, 5—both.

into space—and it could happen anywhere, at any time, though according to Jason Phipps Morgan, that may be sooner rather than later: Northern Eurasia is just starting to rift, and with the immense pressure built up beneath the Siberian Craton ever increasing, the right preconditions exist for a catastrophic Verneshot event to occur. The Yellowstone Caldera, as well, could present some signs of a potential Verneshot . . . that is, if the impending supervolcanic eruption doesn't vent the pressure first. And when a life-destroying volcanic eruption of unseen proportions is your best-case scenario, you're pretty well fucked. Russia and the United States engaged in a race to see who gets to space first; this is like the Cold War all over again, except it's not so "cold" this time and "getting to space" would be done in tiny gooey pieces.

But hey, look on the bright side: You could literally shoot an entire country in the face with 100 percent pure America. That'll show all those pinko commie terrorists, assuming that they don't fire themselves at you first.

14.
POLE SHIFT

SOMETIME IN THE very near future, the Earth's magnetic field *will* reverse. North will literally become south, and the consequences of this shift are not yet fully known. Experts cite potential side effects that range from the lovely sounding "northern lights in Cuba" to the slightly less endearing "everybody gets cancer and the Earth spits you into space."

But while nobody is really certain what will happen when it comes, what is sure is its imminent arrival: Earth's poles reverse every quarter million years or so, and it's been around 700,000 years since the last one. Like an unwed, pregnant, teenage library book, we are both long overdue and seriously fucked. The Earth's magnetic field is caused by the rotation of the molten metal in the upper core, just below the Earth's surface. That magnetic field diverts harmful particles and radiation from space (everything from dust to gamma rays) away from the bulk of the Earth. Most of us probably don't have a solid idea of what the various fields and layers of atmosphere actually do for the Earth; to your average Joe, it's all just so much invisible space magic. That could particularly describe the magnetosphere that surrounds the whole planet and reaches far out into space. The poles, being the strongest points of the field, work like giant magnets, dragging foreign particles toward them and depositing them at the extreme ends of Earth. You can even see this process happening: The light reflecting off of these particles is why we have the aurora borealis, or for those who you hate excessive vowel usage, "the northern lights." Setting aside the somewhat disturbing realization that those pretty green ribbons in the sky are actually a perpetual rain of tiny space bullets, this

Average Joe's Understanding of Atmospheric Layers

Troposphere: Really warm, sandy sphere.
Stratosphere: Planes go here, also guitars.
Mesosphere: Me so sphere-y?
Thermosphere: Where nuclear power comes from.
Exosphere: A sphere with its bones on the outside like a bug.

means that the purpose of the pole is to protect the rest of the world from harm at the expense of the northern and southern extremes. Considering that the southern extreme is Antarctica, which is sparsely populated at best (save for adorable penguins, but do try to avoid thinking about radiated particles giving cancer to baby penguins; it's just too sad to fully contemplate), and considering that—according to the stereotypes that make up my entire base of knowledge—the northern extremes are exclusively populated by Canadians, Eskimos, and Santa Claus, depositing all this space death at the poles isn't really affecting anybody that matters (sorry, Santa).

But that magnetic field is long overdue for a flip, and that could mean some very bad things for us. While the shift is happening, the protection offered by the magnetic field will be drastically weakened—if not absent entirely—during a process known as a "fade." And with that field gone, expect cancer and mutation rates to rise dramatically. If it helps, think of the magnetic field as a sort of space sunblock . . . except instead of shielding your pasty ass at the beach, it shields the entire planet, and instead of getting sunburned if it fails, you get supercancer and flipper children.

Groups I Owe an Apology to After That Joke

- Eskimos
- Canadians
- Children
- PETA (unrelated)

Thus far this has been a "wouldn't that suck if that happened" scenario and, though the timing dictates that it *should* be happening soon, that "soon" could well be anytime over the span of the next thousand years. The odds of the shift happening within your, or your children's, or your children's children's lifetimes has to be pretty low, right? Living through something as rare as this is so unlikely it would be like winning a particularly shitty cosmic lottery of death. But then, this is a book about the end of the world, so you probably see where I'm going with this. . . .

That's right! You've won the cosmic shit-death lotto!

The magnetic field has been fading at an ever-increasing pace for the last three hundred years or so, and right now it's already down to about three-quarter strength. Just judging by the numbers so far, even if it continues at this rate, it shouldn't really matter to you: We should all still have a few hundred years minimum before it's low enough to affect life on Earth. But that mentality, in addition to being relatively dickish to your great-great-grandchildren, also isn't entirely accurate. We can't rely on the rate of change to be slow and steady. There's an anomaly that exists right now in the south Atlantic called the South Atlantic Anomaly (because they're scientists, not advertising executives; sexy names ain't in their repertoire) that has already started the shift. It is a huge chunk of Earth beneath the ocean where the field is not just absent, but actively switching its magnetic polarity. Apart from fucking up all the mermaids' computers, this particular anomaly isn't going to really do anything, but it does confirm that the shift is going to happen in pieces, and in a nonlinear fashion.

So the rate the field is decreasing has been roughly mapped out. It will be completely gone—as in everywhere—in less than a thousand years, but vast swaths of the planet will also be losing their field in chunks well before that time. But even the areas with still-functioning magnetic fields will be operating at partial strength, and partial protection is just that: incomplete. Every percentage point that the field weakens makes it just a little more likely that you'll be catching tiny molecular comets of radiation with your major organs. So instead of having full protection, like a Kevlar vest, right now you have a Kevlar tube top: spotty coverage at best. Soon, you'll be down to a Kevlar bra as the field drops further in strength, and then before you know it all you're left with is Kevlar pasties: It's still protection, technically, but unless your nipples have some serious enemies, it may not be where you need it, when you need it. It's a pan-global game of strip poker against space cancer, and the game is rigged against you.

While the more reasonable theories argue that the flip is a long and intricate process taking thousands of years to fully complete, a few people insist that the only way a magnetic shift this large can occur is from a huge planetary body passing so close to the Earth that the torque from its influence literally shifts the geographical locations of the poles—sending Sweden to South Africa and vice versa. The continents themselves would shift on the liquid base supporting them, sliding like gargantuan air hockey pucks across the molten interior of the globe. Obviously, this would be catastrophic. It would be exactly like pulling the rug out from underneath somebody . . . if by "rug" you mean planet, and by "somebody" you mean everybody. We would hurl across entire hemispheres at the speed of sound, or be spit out into space like watermelon seeds . . . if watermelon seeds were capable of feeling intense terror and incredibly brief but fantastic confusion. Several notable psychics and New Age philosophers have said that this scenario is "incredibly likely in the very near future," while respected scientists and geologists have gone on record as saying that this situation is "so retarded" that it has actually "cost them valuable IQ points just hearing about it."

Real Science's Opinion on the "Geological Pole Shift" Theory

- Poorly outlined
- Not plausible
- Not possible
- Not unretarded
- Fuck that noise

But the New Age dipshits do have a little something right: They cite the timeline for this event around the year 2012. As mentioned in the introduction to this book, 2012 is the last, best hope for apocalyptic fanboys and fangirls out there just jonesing for some rapture. This is on account of the Mayan calendar, one of the most bafflingly accurate and intricate calendar systems of the ancient world, which ends abruptly in the year 2012. The Mayans believed that the current "phase" of life will end in 2012, and that nothing will ever be the same afterward. Now, as apocalyptic prophecies go, that particular foretelling of the end-times is downright flirtatious when compared to apocalypses prophesied by the other major religions. From Christianity to Scientology, most major religious texts set up the end of the world like it's an opportunity for a special effects bonanza: Blood, lightning, explosions, plagues, demons, spaceships; it's like an '80s metal video come to life, and God is going to play a bitchin' solo before he trashes his instruments—those instruments being all life on Earth.

So while some interpret it as the end of the world, others believe it's merely the end of the world as we know it (see: Michael Stipe). It could just herald a new era of knowledge and peace unforeseen in the history of mankind, and that's what the Mayans meant by "the end of the current cycle of life." Some people—good, hopeful, optimistic people—would agree with that sentiment wholeheartedly. You know who doesn't believe that? Science. For once, science and religion agree; if anything big does go down in 2012, it's probably not going to be an interplanetary group hug.

It just so happens that 2012 very neatly syncs up with the sun's own regular pole shift. Like clockwork, the sun's magnetic poles shift every eleven years, peaking with an increase in electromagnetic solar storms, and if you were paying attention at the time, you'll remember these periods as being responsible for everything from mysterious fires to fucking with your television reception. But if a sun flip occurred when the Earth's magnetic field was weakening, the energy released by the sun could act like a catalyst, thus kick-starting the process.

It's comforting to operate under the assumption that pole shifts take thousands of years to complete, but other evidence has recently been turned up to the contrary. Samples taken from the Steens Mountain lava beds in Oregon indicate that at the time those lava flows were active, the geomagnetic poles were moving by as much as 8 degrees a day! If it kept up at that rate, the poles could've completely flipped in a month. Still others believe that the field does more than turn over in chunks, as was indicated by the South Atlantic Anomaly. They think that for thousands of years during the transition, the bipolar nature of the Earth actually splits into several rogue, roving poles that wander at random across the surface of the Earth. Eight norths!

Things You Want Kick-Started

- Motorcycles
- Soccer games
- My heart

Things You Do Not Want Kick-Started

- Hungry lions
- Your crotch
- The Earth

Moving souths! Navigation will be useless and constantly changing! One day you'll wake up in North America, the next day South! Brazilians will enjoy Budweiser while Wisconsinites guzzle Fanta by the case! *Anarchy!*

And finally, one of the more credible theories for pole shift revolves around something called the Jupiter Wind. Though it sounds like a particularly hackish John Grisham novel, it refers to the influence of Jupiter's "winds" on the Earth's magnetic field. Astrophysicist Frances Bagenal has shown that "magnetospheres of rapidly rotating planets with strong magnetic fields [e.g., Jupiter and Saturn] are dominated by rotation, while the solar wind controls the plasma flow in smaller magnetospheres of slowly rotating planets [e.g., Earth]." This just means that Earth's magnetic field isn't large enough to be completely independent, so a huge factor of its stability is outside influence, namely the sun.

And here comes that troublesome date again, 2012: Not only is it the next cycle of solar magnetic activity, but it also takes place in a window where Jupiter, the Earth, and the sun are all in perfect alignment. Jupiter's self-generated electromagnetic field is normally pushed into the far reaches of the solar system by the constant solar wind; with Earth blocking the sun's influence, Jupiter's magnetic field would respond to the lack of pressure from the sudden absence of solar winds by sending a massive surge of its electromagnetism in the direction in which that pressure was blocked. That's Earth.

Think of it like this: You're a skinny asthmatic nerd child (with a weakened magnetic field), already extremely prone to beatings on the best of days, but today you've accidentally picked a fight with the biggest, baddest motherfucker on the playground (that's Jupiter). As long as the teacher is watching, exerting an external pressure on both of you, you both keep to yourselves. Your magnetic field (in this analogy that's your face) remains intact only because the teacher is suppressing the bad motherfucker. Now the teacher leaves for a smoke break, creating a sudden lack of restraint in the behemoth, and he's pointing right at you . . . with his fist (or the Jupiter Wind).

Now we have a space tsunami hitting a weakened magnetic field from one side at the same time that the Earth is under assault by the powerful solar winds generated by the sun's own pole shift on the other side. It's a perfect storm that batters the Earth with unbelievable force and could potentially wipe out our planet's weakened magnetic field instantly. So let's go back to our metaphor one last time: You're a weak, feeble child, and the teacher is nowhere to be seen. What is to be seen, however, is the giant fist (or electromagnetic wave) hurtling at your face. You cannot dodge this fist; you can only take the hit and all of its consequences. That would be bad enough, but hey, you'll live, right? Well, no, because this time a miracle happens: This time an errant delivery van (in the form of solar storms) comes careening through the playground fence and heads straight for the bully. As the bully swings his fist at you, the van slams full speed into the elbow of his punching arm, sending the already devastating blow at you with unimaginable force (both solar storms and Jupiter winds hitting simultaneously). Due to freak coincidence, you're no longer just getting punched; now you're getting punched with a truck.

Now that we know we can't avoid it, we should at least know some of the expected consequences. For starters, because a pole shift can happen in stages, we could have a series of temporary magnetic poles wandering the Earth at random.

So not only would people have no idea where north is, but even more confusingly, there could be multiple norths. Obviously this would play havoc with all modern navigation, and anything relying on compass directions would be little better than a confused, lost male stereotype, unable to ask directions even if they wanted to.

This means that all migrating animals would also lose their direction, which could lead to such wacky scenarios as Canada geese migrating to the Bahamas for the summer! And the subsequently less wacky following side effects of that migration:

The total extinction of Canada geese.

Reversal of the Earth's magnetic field would also affect the electrical conduction of the Earth's molten outer core, which in turn might lead to heretofore unseen volcanic activity. It wouldn't necessarily lead to supervolcanoes, but everything that might be about to erupt would immediately do so. The shift could also act as a catalyst for earthquake activity, but really, at this point? It's just showing off. Everybody's already on fire or flying their planes into the ground. You don't have to be a dick about it, geomagnetic field.

But hell, even if we survive the Earth self-destructing, the magnetic field can have serious effects on the human body all by its lonesome.

With our atmospheric shield down, all the harmful space particles are allowed through at full strength. Gamma rays, cosmic rays, and solar radiation will all come blasting right in and—aside from sounding like

The Geomagnetic Field

Kicking you while you're down, then putting you down lower while continuing to kick you, then burying you and digging you back up just to kick you some more.

ammunition for supervillain weaponry—all of these particles have one thing in common: They're potentially deadly. They can cause genetic mutation, cell death, and cancer in any and every little thing, just by their mere presence. So when the pole shift happens, we will no longer have any defenses left against the cancer-causing, DNA-mutating particles from outer space spawned by the cosmic forces plotting our demise. OK, so that may be a bit overblown; space doesn't necessarily *want* you dead, but it will kill you regardless. At least you can pretend it's got a motive, and maybe give your impending murder a little meaning.

And remember those wacky Canada geese who basically lost themselves out of existence? They did that because they have extremely small amounts of a mineral called magnetite that syncs their brains with the magnetic field, thus giving them their sense of direction. What other animals have this same direction-giving mineral in their brains? Why, you do! One researcher, focused on the effects a changing magnetic field can have on the human body, found that by varying the strength of or entirely removing the magnetic fields from her subjects, serious coordination problems arose. The subjects immediately became disoriented and clumsy, falling down more easily. They couldn't even complete simple body tasks, like touching their fingers to their noses or standing on one foot. Some became motion sick, and all reported serious difficulty concentrating.

The decreasing magnetic field, the one we're living with now, could very well have similar, if not more profound, effects. After all, the subjects in this study were subjected to weakened fields for only a short time, whereas the real pole shift could take hundreds of years to complete. How many generations of a stumbling, half-retarded human population tripping comically down stairways can occur before we, as a species, just cease to exist? It could be death for us all. Death, by mass pratfall. But hey, on the plus side, if there is an America left after all of this, *America's Funniest Home Videos* will have material until the end of time.

BIOTECH THREATS

The fear of an apocalypse triggered by biotechnology is perhaps the most immediate concern out there: Genetic manipulation, biological weapons, and superdrugs set off the panic response like little else, and the risk seems to be becoming more real by the day. But what would need to happen for the worst-case scenario — a worldwide, man-made biological apocalypse — to actually come to pass?

First, you would need incentive, some benefit tempting enough for the masses to risk the very fibers of their being by introducing manipulated genes into their own systems. You're probably not going to be game for freebasing possum blood unless you get something pretty cool in return. Free cable, at the least.

Second, you would need contagion, a way for these genes to spread from person to person, or otherwise propagate. While it may suck for you, personally, to have your mind taken over by telepathic heroin, if it doesn't affect everybody it's just back-page news.

Finally, you would need lethality: Because, well, it's just not the apocalypse unless everybody dies.

15.
BIOTECH INCENTIVE

AS PREVIOUSLY MENTIONED, humanity is not entirely self-destructive. We won't put ourselves in jeopardy for no reason. And we're not going to! The good news about the incentive phase of a biotech apocalypse? You're going to be able to do some pretty cool shit.

The most alluring field of biotechnology right now is that of athletics, and for good reason: Athletes have been looking for new and undetectable ways to cheat since ancient times, when the very first Olympic wrestler greased himself up before a match. (Greek wrestling, much like its modern contemporary, professional wrestling, was the most socially acceptable way for butch men to get naked and grope each other with lubricant.) And ever since that first naked, cheating man-orgy, we've taken a big cue from their example. Genetic experimentation owes a great debt of gratitude to brave, unethical souls like those ancient Greeks, whose modern equivalents are competitors willing to risk shriveled testicles and crippling bacne just for a leg up on the competition.

Activities Marginally Less Homoerotic Than Wrestling:

- Carnaval
- Naked disco
- Hard-core gay sex with a bunch of other dudes (all the time)

A chief problem is that of "gene doping." It's the biological equivalent of pimping your ride, and it will soup you up in every way short of painting flames on your chest and installing monitors in your ass. In general, the term "gene doping" refers to any modification of the human body through enhanced DNA. It's used for a variety of therapies, but in the field of professional sports there are some specifically desired uses: Injecting extra copies of naturally occurring genes, like those that create endorphins, might render you essentially immune to pain and fatigue. Or perhaps you'd rather have some extra testosterone, making you larger, stronger, and substantially more entertained by NASCAR. Before the prospect of gene doping, there was simple blood doping: Some athletes prefer injecting themselves with erythropoietin, or EPO, to boost their own red blood cell count. This enables better, faster, stronger oxygenation of the blood and therefore better stamina, fuller lung capacity, and less recovery time. But in the realm of modern science, that's practically caveman shit. Why, all the way back in 1998, an entire cycling team in the Tour de France was expelled for EPO abuse, despite the fact that it's nearly impossible to trace, because evidence of almost any kind of gene doping can be explained in other ways: There are disorders, birth defects, or just different genetic profiles that could be responsible for unusual genes, thus allowing any athlete to dismiss accusations of cheating by attributing it to a tragic birth defect that rendered him substantially more awesome than the average man. With gene-doping technology, we're looking at the same essential effects as blood doping; we're just going to make them permanent. By altering DNA to increase the production of EPO in athletes' bodies, they no longer need to inject untraceable superdrugs to up their performance;

Beware: EPO can easily be confused with ELO in casual conversation.

If you find yourself struggling to tell the difference, here are some tips to help you differentiate between the two.

- EPO can also be called "blood doping."
- ELO can also be called "dope" by people who own BeDazzlers.
- EPO is used to better performance.
- ELO will never be followed by the term "better performance," unless it's meant ironically.
- EPO is an illegal substance, and very hard to come by.
- ELO is not yet illegal, and can be found in yard sales across the country in the box marked FREE CRAP, next to broken He-Man toys and a single ski boot.

they're already producing it. You've got little drug cartels up and running in your bloodstream.

Consider vascular endothelial growth factor (VEGF). It covers the other end of the spectrum from EPO's stamina-enhancing effects by stimulating the growth of new blood vessels, increasing blood flow, thus improving burst-style muscular strength. Its intended use is for treating muscular diseases, allowing patients with conditions that constrict their blood vessels to regain their muscle mass and, most likely, will eventually wind up helping old white men get boners—because pretty much all drugs end up in that shriveled-boner category. The thing so devious about VEGF is that, even if you manage to track this one, the gene that produces it is delivered to cells in the body in the first place by piggybacking on the common cold virus. So finding modified VEGF genes in the herculean superman who just hurled the Lithuanian swim team out of the coliseum will at best just prove that he had a case of the sniffles at the time. Which, if anything, just makes him look all the more impressive, doesn't it?

But that's only for the obsessed athlete, right? How could this possibly benefit the common man?

Well, for one, consider the elderly: Typically regarded as frail and easily knocked around for one's amusement when you're desperate to compensate for your microscopic manhood (not that I would know anything about that), their weak constitution is largely due to the natural loss of muscle mass. That's just genetics, of course, and thanks to advances in biotechnology, there are experimental drugs in testing right now that help restore that lost muscle. They could restore enough, some say, to even equal those levels found in healthy young adults. That's not just halting one of the worst aspects of aging; it's turning it backward! The drug is called MK-677, and so far it's been proven safe, effective, and able to restore up to 20 percent of total muscle mass in the human body. That's a full fifth of the total human potential, effectively taking senior citizens back to the same overall body strength as a healthy twenty-year-old, according to Michael O. Thorner, a professor of internal medicine at the University of Virginia Health System.

That's a real thing now: geriatrics with the bodies of twenty-year-olds. That's not science fiction. It'll probably be commonly prescribed well within your lifetime if you're reading this book (because hey, let's face it, the elderly are probably not the key demographic for a book full of dick jokes about cutting-edge science).

So you're already one-fifth immortal and possess nigh superhuman stamina just by taking pills, but wait! There's more!

If you call now, the future will throw in super strength, absolutely free!

These deals are so crazy they'll stuff and mount their dead mother!

Scientists have discovered that disabling a protein called myostatin can

In case the elderly are, in fact, enjoying this book right now, please accept my most humble apologies and my thanks for your readership. To make it up to you, the following text is exclusively for your enjoyment:

- What's with this Twitter thing? In my day that was called a sentence.
- Kids will stay off your lawn if you sprinkle it with cayenne pepper. That'll give 'em something to cry about!
- Why don't more places serve rice pudding?

double the overall size of musculature in most mammals. This has only been verifiably proven in mice so far, but theoretically it should work on virtually any mammal. There's a naturally occurring condition in several species that's quite similar, where the animal's myostatin is genetically suppressed. Whippets—those malnourished junkie-looking dog-rats—can suffer from a hypermusculature disease that creates "bully whippets": Genetic freaks that are gargantuan, absurdly muscled even without exercise, and overall resemble nothing so much as a canine Incredible Hulk. It's not a debilitating condition, either—that muscle mass is fully functional. Bully whippets have double the strength of their normal counterparts, and as a result can run at double the speed. There's even been one recorded case of this condition occurring in a human. A German boy was recently diagnosed with the disorder. He's been monitored from the day it was first discovered he had the condition, and all subsequent tests have verified that he is indeed in perfect health. This means that, theoretically, an artificially induced muscle-doubling disorder should also be entirely safe. Of course, the second professional weightlifters heard of this development—a procedure that could render them roughly twice the terrifying abominations of nature that they already are—the researchers were swamped with offers to volunteer as test subjects. It's not like the appeal of strength in a pill is limited to the Buick-resembling man-monster demographic either. Se-Jin Lee, one of the professors at Johns Hopkins University responsible for this discovery, states that every major pharmaceutical company is working on a myostatin blocker for treatment of diseases like muscular dystrophy.

If it's already being tested medicinally, the demands of

the free market (read: anybody who wants totally kickin' abs for doing absolutely nothing, which, by last count, is everybody) dictate that it will soon be available for more casual usage as well, within just a few years by some estimates.

If strength doesn't appeal to you, what about stamina? There are biotech endurance enhancements in the works as well: Scientists at Case Western Reserve University genetically altered mouse embryos to limit the muscle response for burst-style energy. The effect of this was the lessened potential for energy spurts, but massively increased endurance. Some other little side effects were seen, like insatiable hunger, tripled life span, and a tremendously increased libido, along with a "very, very aggressive nature." In other words, these are furious, untiring, and horny-as-hell mice . . . possessing a *ravenous, unnatural hunger*. The researchers, as they did with VEGF, EPO, and MK-667, freely admit that there is also the potential for abuse of this as a performance enhancer in humans, but they do strongly warn that these same potential side effects might affect humans the same as they do mice. So maybe take a rain check on that workout in a pill for now—side effects may be nausea, headache, stiff joints, endless

Sure, zombie drugs may be disturbing, but consider the side effects of some already available pharmaceuticals:

- Sleep sex (Ambien)
- Impulse gambling (Mirapex)
- Necrotic flesh (Provigil)
- Old-guy boners (Cialis)

rage, consuming lust, insatiable hunger, and other such *general fucking zombification*.

So there are zombie pills! Wonderful! But surely you didn't think that was all?!

No! There's even more! Science is *clinically retarded for deals*!

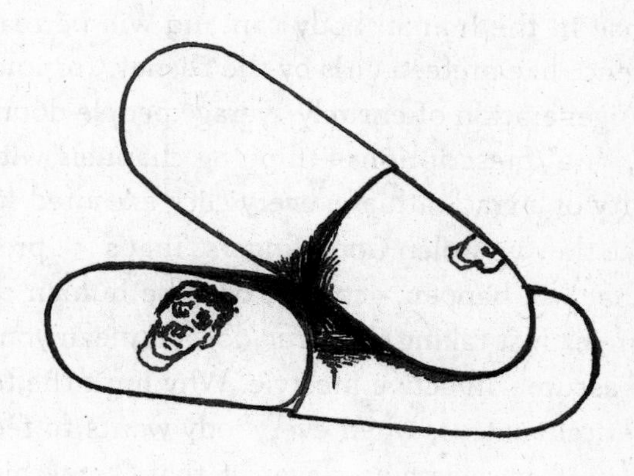

You've heard the phrase "in the zone"? The mental, spiritual, and physical state where you feel you can simply do no wrong—from shooting a fine game of pool to breaking world records—being in the zone is the common moniker for what scientists refer to as a "flow state," a feeling resulting from a dopamine surge accompanied by a few neurochemicals, such as serotonin and norepinephrine, which generally serves to decrease reaction time, and even alter our perception of time itself.

That's in a pill, too: You'll be able to be "in the zone" within the decade just by popping some anandamide analogues. Anandamide, first identified back in 2004 by Georgia Tech neuroscientist Arne Dietrich, is the chemical responsible for triggering these flow-state highs and, as you've probably learned by now, pretty much any naturally occurring response in the human body can and will be manipulated by science like preteen girls by the Disney Corporation. Picture it: A generation of entirely average people doping up on "in the zone" prescriptions—flipping channels with the mental acuity of a crack athlete, every click executed so perfectly it's like they've stolen God's fingers. That's . . . probably the most that will happen, actually, because human nature doesn't change. Just taking this drug doesn't mean you'll automatically assume an active lifestyle. Why limit the feeling to the athletically gifted, when everybody wants to feel like they're the best at *something*—even if that "something" is just the Hundred-Ounce Pee Break or the Triple-Digit Channel Change?

So we've established that there are real projects already under way with proven results that give humans everything from effort-free workouts to improved mental faculties. With kickin' abs, endless stamina, mental acuity, and nigh-spiritual flow states available in pill form in under a decade, it's not a matter of if you introduce biotech to your system, but how much you can afford and how awesome you want to be that day.

When you consider that we're the generation that

bought the Thighmaster and the Ab Cruncher, it's safe to say we'll probably make some room in the budget for buff-untiring-genius-time-manipulator-in-a-pill. And that's how it starts; the massive incentive for biotech makes it a household name. Soon you won't think anything of modifying your genetics on the fly, and biotech will be everywhere.

16.
BIOTECH CONTAGION

AH, BUT NOW COMES the downside: The contagion period. For nearly every experiment with beneficial discoveries, there's one with horrific, lethal mistakes. These advances aren't coming out of thin air; they're developed in lab animals —mice, for the most part. The beneficial drugs are tested, tweaked, and refined in lab animals, and then, if approved, eventually synthesized for humans. Supposedly, this method ensures that there's no danger of something unintended crossing over to us from the labs, because the mice and other animals used for experiments are somewhat genetically isolated from us. Even if we do accidentally unleash a plague in the pursuit of the betterment of man, it'll just wipe out some lab rats with nary a vegan to even shed a tear. But for any failed experiments to actually pose any danger to humanity there would have to be more common genetic markers between the lab animals used in these early testing stages and the common man. Like, say, if mice could shoot human ejaculate or something equally horrifying.

Oh wait, they already do exactly that. Surprise! Science hates you.

If Science Doesn't Hate You, Explain:

- Styrofoam
- The atom bomb
- Roughly 70 percent of this book

This, shall we say, *sticky situation* (and we probably shouldn't) all started with researchers at the universities of Pennsylvania and California, who initially just managed to provoke mice into growing viable monkey sperm. While "provoking mice into growing monkey sperm" brings to mind some rather disturbing images of furious primates and sexually antagonized mice, it was all pretty innocent. Perhaps they just wanted to give the internet even *more* bizarre pornography, or perhaps they were just out to instill some confusing urges in horny rodents for the sake of comedy, but most likely the scientists were hoping to help endangered species by synthesizing sperm on a large scale.

That's what their proposal states is the ultimate goal, anyway, and why lie about something that embarrassing? If it wasn't true, a more flattering lie than "Professional Rat Molester" should have been easy to come by. So while the researchers in Pennsylvania may have started all of this, it was Canadian researchers hot on their tails (which takes on a rather gross subtext in this instance) who took the concept even further, instilling the ability to produce human proteins in mouse sperm. The Canadian mice now ejaculate human growth hormone—a fact that cannot be doing the tourism industry any favors.

To create this monstrous hybrid of mad science and pornography, the researchers first started by targeting a DNA sequence present only in male sex glands. Once the gene responsible for sperm production was identified, they just spliced in a sequence responsible for human growth hormone . . . and that was it! Apparently nature's pretty flexible about this shit. You want mice to come like people?

No problem, says nature!

She's in a giving mood this millennium. What else you want? Some crocodiles with sonar, maybe a feathered dog, hey—how about some gorillas with scales? Genetic splicing is apparently limited only by your imagination, funding, and willingness to fight off snake apes. The Canadian researchers at least had good reason for their spunk meddling; their process has some impressive potential for the pharmaceutical industry. Insulin is already being "farmed" in animals

Great Suggestions for New Canadian Mottos

- "Come to Canada, where politeness is a prerequisite!"
- "Visit Canada, we have the Internet now!"
- ~~"Vacation in Canada, our mice cum like people!"~~

for human use, but differences in chemical composition results in varying degrees of effectiveness, depending on the animal. By altering these mice to produce genetically identical human insulin in their sperm, the Canadian scientists expect to have a new, completely safe, and entirely effective method for producing high-quality medicine on demand.

Taking a money shot in the veins from Mickey once in a while is a small price to pay for that kind of progress.

But the researchers soon realized that tiny mice have tiny mouse orgasms—not producing a lot of the desired substance—and instead moved on to re-create their experiment in larger animals, eventually settling on boars and cattle, which can produce more than 300 ml of semen up to three times a week on average. Though these HGH experiments are being performed only with semen for the time being (because if something's worth doing, it's worth doing grossly), eventually the hope is to encode the modified gene sequence so the hormone is produced in more amenable substances like cow's or goat's milk . . .

Or pig urine!

No kidding. That is a specific goal. The researchers involved in these experiments hoped to farm something slightly less horrific than mutant mouse semen (and really, who could blame them?) and so began looking toward replicating the results in other fluids. And either some guy just totally missed the point of this exercise, or else had some really unsettling sexual priorities:

Scientist 1:
Does it have to be rat jizz? I mean, sure we're extracting the pure chemical from the substance, and using the end product isn't like letting a mouse pop one off into your veins, but it's still pretty weird. I'm not sure how people will feel about this.

Scientist 2:
Well, what if we splice it into a more socially acceptable fluid?

Scientist 1:
Yeah, yeah! Like blood or, hell, even milk!

Scientist 2:
Or pig piss!

Scientist 1:
Yeah, I . . . wait, what?

Scientist 2:
Think about it! You could drink all the pig piss you wanted, and it would even be good for you!

Scientist 1:

I hate everything about working with you.

The only potential downside to using something like milk would be the waiting period: Not only would you have to wait until the genetically modified animals reach sexual maturity to begin farming, but you'd have to wait for them to lactate, too. Honestly, though, we as consumers already have to wait that long for normal dairy products, I think we can wait a few extra weeks to avoid mainlining mouse man-batter.

Another key concern of genetic experimentation in general is that of containment viability. Modified genes tend to carry over naturally, after all, and it takes only one pair of star-crossed cow lovers and a broken section of fence to introduce the traits to other cattle. Cattle that have other uses. Cattle that produce milk.

Cattle that you eat.

Methods of Ensuring Your Meat Is Safe (In Order of Difficulty)

- Burn everything to ash.
- Consume only products you grow yourself.
- Stop eating meat (this includes bacon).

If this does happen, you would not only have a nearly untraceable dose of pharmaceutical-grade drugs in the food supply, like that aforementioned insulin, but that would also further narrow the gap between those previously isolated laboratory animals and humans. Perhaps this interbreeding leads to medicinally fortified foodstuffs, or perhaps it leads to a Grilled Ham and Insulin sandwich that, while undoubtedly delicious, is unfortunately always served with a heaping side of hypoglycemic coma. That example is just to illustrate a point, though; the real problem is much larger.

While consuming genetically modified animals won't alter our own genetic code, the fact that our livestock is now one step closer to humanity opens a gateway to a multitude of diseases and mutations that can cross over between species, not to mention some potential for unforeseen changes to the source animal. So at first you just have lab mice ejaculating human protein, but what if that starts to carry over in reproduction? If just one animal gets out, natural selection would start to take hold. The altered gene might become hereditary, and then we all get to find out exactly what effects a human gene in the mouse population can have. What's your wager? Is the end result the walking incarnation of hugs, like a real-life Mickey Mouse, or something more akin to the drugged-out psychoses of the Rats of NIMH? Is human growth hormone too obscure a worry to illustrate this danger? Well, try this horror hat on for size and tells me how it fits:

German scientists have successfully modified a batch of mice with a human gene for speech. Now, I know what you're thinking:

"What? Fuck that."

But it's true! In an effort to better understand the evolution of human language, they've spliced a gene called FOXP2 into some lab mice. Humans with a low FOXP2 count can suffer from anything from speech disorders to cognitive inhibition, and though a relative of this

Most Popular Mouse Phrases (If Mice Could Speak)

- Foodfoodfoodfood.
- I like climbing!
- I poop where I stand!

gene is present in all sorts of animals, the strain in humans is entirely unique to our species.

Well, it used to be, anyway.

What did we talk about earlier in this chapter? Remember how amenable nature seems to be to changes in genetic structure? If you want talking mice, nature is probably quite happy to oblige.

There's actually evidence of change happening already: The enhanced German mice showed drastically increased nerve activity in the language center of their brains, for one. They can't speak or anything yet—this shit is crazy, not retarded—and even if the language centers were amplified to human levels, that doesn't mean anything like human speech would develop. But it does mean that they're now better at communicating with one another. That's a clear biological improvement. In terms of evolution, that means it's likely to favor a gene and pass it on to future generations. That's not idle speculation, either; the scientists in charge of this experiment freely admit that, though this research was initiated just to study the evolution of *human* speech, it could well give these mice a push down that same evolutionary path.

If there's one thing we don't want communicating and coordinating with one another, it's rodents. Disease-spreading creatures that have already, at least once in history, participated in the near total destruction of the human race, thanks to their role in spreading the Black Death. And that was without boosted language skills and human sperm! Now they can not only jizz disease on your face, but ask you who your daddy is while doing it.

Isn't progress wonderful?

17.

BIOTECH LETHALITY

SOME OF THE WORLD'S most devastating diseases, from the H1N1 "swine flu" to SARS, have made the jump from animal to man naturally. Though transspecies diseases have fortunately been relatively low mortality in recent years, new variants of the flu have historically been potential extinction-level events. Consider the Spanish flu of

1918, whose total estimated death count varies from 30 to 50 million. That's more fatalities within a year than in most of America's major wars of the last century. It actually was a kind of swine flu; it jumped from pigs to humans and eventually wiped out the equivalent of a large country. Talk about your rags-to-riches story! We're now more reliant on livestock than ever, and it's all conglomerated—roughly 80 percent of America's beef is currently processed by less than half a dozen companies, and a full third of the entire country's milk comes from one single solitary company. You can easily see why an outbreak in a food processing factory, in today's world, could kill on a much larger scale than it could've a century ago. Thanks to corporate rule, we not only wear the same brand of T-shirts, buy the same brand of coffee, and watch the same TV shows, but we also share a common pool of food products. And that's where we drink all of our milk from!

Out of the contagion pool!

According to the above numbers, if a major dairy supply were to be contaminated, there's a one in three chance that you'd end up drinking it. So here's a worrying thought: We have security measures at airports and train stations . . . but how well-protected and monitored are the dairies? Terrorists got pilots' licenses; I'm betting that they can get jobs as milk skimmers at the competitive rate of three fifty an hour and half a pound of free cheese per meal break.

To make matters worse, not only is there contamination in our food

Less Disturbing Pools Than the "Contagion Pool"

- Unchlorinated swingers' hot tub
- Disused hobo bathhouse
- Office pool on estimated time of coworker's newborn baby suffering crib death.

distribution, the actual livestock themselves aren't much better. Thanks to the overuse of antibiotics in feed animals, superresistant diseases are rapidly on the rise. That cow whose sole purpose in life is to turn its ass into delicious cheeseburgers isn't exactly going to get primary care; farmers have chosen to mass dose their herds with antibiotics and hope to catch what they can. This not only leads to a reduced efficiency in the cows' immune systems, but in our species' as well. That cow's last meal probably contains those same antibiotics (I mean, what's he gonna order, steak?) and that makes it *your* current meal, if you then eat said cow. So human antibiotics have an overall reduction in efficiency because of mass-medicated cattle, and that sucks (fuck you, cows—no wonder we eat you), but Dr. Arjun Srinivasan, an epidemiologist with the CDC, actually thinks that, just considering the sheer amount of antibiotics dumped into the food supply, antibiotic overdose in animals may actually endanger the human race more than any potential overdose in the humans themselves.

But this is all so nebulous: Biotech "affecting the immune systems of cattle" isn't exactly a sexy apocalypse. So let's consider the actual effects one of these animal-spread pathogens could take. For example, *Toxoplasma gondii*. It lives most of its life in rats, but can only mature to adulthood in the belly of a feline. When it's ready, it causes the host rat to seek out and loiter around cat urine—because where there's cat piss, there are cats—and, when the rat is inevitably eaten, the parasite is successfully transferred to the feline host. *Toxoplasma gondii* is an organism that possesses its host and fosters a death wish, because it can thrive only if the host dies. But fuck those rats, anyway. That couldn't possibly affect you, right?

Well, actually, half of the entire human population is infected with *Toxoplasma gondii.* Not "throughout history." Not just "at risk." But literally and presently fully infected.

Right.

Fucking.

Now.

This is particularly bad, because severe cases of toxoplasmosis manifest as full-blown paranoid schizophrenia: voices, delusions, hysteria—the works. Toxoplasmosis is the reason that pregnant women aren't supposed to handle cat litter, why newborn babies aren't supposed to be around cats, and though there's no actual basis to believe it, it certainly goes a long way toward explaining crazy cat-collecting ladies if they're actually infected by mind-controlling parasites harbored within the cats themselves. Even in minor cases in people whose immune systems aren't compromised (fifty/fifty chance says that's you), the majority of infected males suffer from neuroses, guilt, and nervousness from even mild cases, while infected females are more aggressive and outgoing . . . and have drastically heightened sex drives. That explains why those cat ladies put out so easy, am I right, fellas? Hell yeah! Maybe our parasites can get together and high-five later.

It's basically a viral parasite that transforms women into ballcrushers and men into pussies. That's a little unsettling, but hey, gender roles are different in modern times anyway. No harm, no foul. Until you consider that these side effects could be

Arguments Against Owning a Cat

- They pee on everything.
- They don't really do tricks.
- They don't really like you.
- They're infested with things that infect your brain and drive you insane.

dangerously exploited, because—like it or not—the modern military remains primarily male. Not hard to see the appeal of a biologically engineered toxoplasma used as a weapon—an army of butch Jack Dempsey hard-asses instantly emasculated into nebbish, Woody Allen–style timidity, too insecure about their manhood to order a steak correctly, much less fire eighteen bullets into the face of a rampaging RoboNazi (hey, it's the future, right? I'm just assuming we'll fight enemies a little cooler and less ambiguous than "brown people who don't live on this continent" by then).

And remember, those are mild cases. In addition to schizophrenia, severe cases manifest side effects like blindness, cerebral palsy, severely diminished coordination, and even death. Toxoplasma infections are mostly kept in check by our immune systems, so severe cases are rare. But then, there are any number of reasons why our immune systems might not be up to snuff, from the serious—such as AIDS or chemotherapy—to something as mundane as the flu. With half of the population already harboring *gondii* parasites, any immune-suppressing assistance at all, and *gondii* could start wiping us all out.

But as terrible as the severe cases can be, we're still alive. If we catch it in time, it might not be an apocalyptic-level event, even *if* superstrains were engineered. But keep in mind that it takes only one mutation for the host roles to change.

If that happens, we could very well become the rats, gestating the parasite until it decides it's time to move on to something better. So maybe tomorrow's the day you wake up loving the distinct musk of lion piss, and it's some lion's turn to get uncontrollably horny and schizo. Don't outright dismiss the likelihood of this kind of mutation, either. Consider that with military funding and advances in modern science, it would be a snap to modify a few strands of DNA and let a modified *gondii* loose. It sure would make the army's job a lot easier, after all, if all of their enemies inexplicably began loving the smell of recently plucked hand grenades.

But hey, guys, look at the bright side: If there's a death wish–inspiring parasite living in all of our brains, at least it's paying the rent in skanks.

So far, we've talked about how biotechnology has not only elevated the threat level of the viruses themselves, but also, through our extensive use of agricultural biotechnology, created a mass contaminate pool larger than any in the history of the human race; we all drink from the same watering hole (technically it's a milk hole, but that's just pornographic), so let's move on from the general dangers and really home in on sweet, sweet genocide: In August 2000, a bacterium named *Klebsiella pneumoniae* (that's right, in the same genus as that *other* world-destroying bacteria) was discovered in New York University's Tisch Hospital. Dr. Roger Wetherbee, a physician there, describes it most succinctly in this excerpt from a *New Yorker* article:

> "It was literally resistant to every meaningful antibiotic that we had." The microbe was sensitive only to a drug called colistin, which had been developed decades earlier and largely abandoned as a systemic treatment, because it can severely damage the kidneys. "So we had this report, and I looked at it and said to myself, 'My God, this is an organism that basically we can't treat.' "

That sounds like dialogue from a bad action movie. Did he dramatically remove his glasses after saying "my God"? Did motherfucking lightning strike when he said "this is an organism that basically we can't treat"? Thankfully, this episode in New York was the only major outbreak of this strain in the United States. But unfortunately, this particular strain of *Klebsiella* had refined itself so extensively in the clinical environment—thriving on weak patients, immunizing itself to antibiotics—that it couldn't even be killed with *industrial disinfectants.*

Industrial disinfectants are hard core; I think that's what finally killed John Wayne; I think that's what finally toppled Communism—hell, that's probably what killed disco! Though a form of bleach *did* eventually kill *Klebsiella pneumoniae*, it was eventually proven resistant to everything from ammonia to phenol. This sounds unkillable, like the goddamn Highlander of bacterium. I think disease just found its first superhero.

Huge precautions were taken, and hospital staff basically donned biohazard suits and flamethrowers to burn the infected before they could be turned. That hospital should have been more sterile than a *Murphy Brown* episode, and yet still the bacteria thrived. Patients started getting bloodstream infections from the bacterium, which is the most lethal infection one can get, before hospital staff eventually contained it. By that time, nearly half of all those infected had died. This was in a space as confined as a university hospital, with strict containment procedures and experience in combating the spread of disease. Imagine if this outbreak had occurred anywhere else? With this high a mortality rate, it could have halved the world's population practically overnight.

If You Liked That *Murphy Brown* Reference, You May Also Enjoy

- Lattes
- Khaki
- Your grandchildren
- Chicks in shoulder pads
- Paisley

Thank God our dairies are so well protected!

But hey, why even develop new microscopic murderers when the classics never go out of style? Scientists found they were recently able to synthesize the polio virus from scratch, presumably in an effort to give us all tiny T-rex arms so we can't fight back when the government Thought Clerics come a-calling. It's been theorized that the same process, a rather simple one by all indications, could also be used to synthetically manufacture similar viruses. But any contender would have to have a simple cell structure like polio, so they still can't do anything *too* complicated, and at least that's somewhat consoling . . . if you stop reading this chapter right now.

Other "simple" viruses in this category include both Ebola and that superlethal Spanish flu from 1918.

That's already happened, actually. Researchers have already synthesized Ebola. They used something called "reverse genetics" (which is presumably just like regular genetics but practiced only on Opposite Day), and presto!

You've done it, scientists!

You've re-created one of the scariest viruses on Earth! Now just hand it over to your boss—the unsettling fellow in the iron face mask and velvet cloak who commutes to the job site every day in a floating castle—and I'm sure he'll be quite responsible with it. He might even promote you!

To an early death.

Shit, are you paying attention, Hollywood? That's how you write a one-liner.

But that's not to say this technique needs a full lab of professionals to replicate. No, all that's needed is a rudimentary set of chemistry tools and the genetic sequence of what you want to replicate. All of that information—the technique, the tools needed, and even the genetic sequence of the

Spanish flu—is, like gaping anuses and furries, just waiting to ruin your life forever . . . on the internet.

So far, we've been talking about accidental by-products or potentially utilized biological weapons, but there's a far more likely way that the biotech apocalypse could happen: completely on accident.

Australian researchers, in an attempt to control the exploding number of wild mice, engineered a variant of mousepox intended to sterilize the population. But they screwed up and inserted one little extra gene, and what was supposed to be just contagious birth control instead became an amazingly lethal plague, with fatality rates approaching 100 percent. The virus spread like wildfire, and the researchers just barely managed to hold it in check. The truly frightening part, however, was the virus' remarkable similarities to human smallpox. It just had a much higher mortality rate and a more aggressive rate of infection.

Other Atrocities That the Internet Is Directly Responsible For

- 2 Girls 1 Cup
- A Database of Plagues
- This book

Now, we can play a tiny funeral dirge for Fieval and all his pals later, because the main factor here is the similarity between mousepox and the human equivalent. As it stands, we have nearly no immune response to the smallpox virus—it was virtually wiped from the planet, so there is no cause to vaccinate against it. However, if it were to return right now, researchers estimate the fatality rate at nearly 20 percent. That's one in five people. And that's regular smallpox. If this engineered mousepox were to cross over, not only could the fatality rate be around 100 percent, but the virus has proven even more contagious than its traditional counterpart. And worst: We have no vaccine for it. Smallpox could be prevented, but because of that one little altered gene, there would be no established defense against the modified mousepox. The researchers assure us that there is no immediate danger from this super mousepox; though the strains are quite similar, it is still impossible for the virus to bridge that gap between human and mice genetic dissimilarities and endanger humanity.

So we were lucky; the little bit of dissimilarity between our DNA made this virus a nonissue for humans . . . but this was all well before the Canadians started screwing around with mouse spunk, of course.

We've established the impetus to willingly expose ourselves to genetically altered materials; the fact that more and more experimentation is resulting in accidentally created, never-before-seen diseases; and, finally, the existence of a new bridge between humanity and these lab animals that the diseases can use to cross over. I think it's officially time to start getting scared . . . and all of that isn't even factoring in the horrifying and now very real potential for accidental pregnancy via supermouse rape. Hey, it could happen. You might think it pretty unlikely that you'll be catching a tiny rodent facial anytime in the near future, but remember, thanks to that endurance experiment, some rodents are now very aggressive, hypersexual, freakishly strong, and untiring. Even if they're not actively out trying to bone humanity, somebody has just significantly upped the odds of you getting caught in a never-ending supermouse orgy.

Right in the path of their deadly plague orgasms.

ROBOT THREATS

Everybody is well aware that robots are out to kill us. Simply take a cursory look at the laundry list of movies—*The Matrix, The Terminator, 2001: A Space Odyssey, Short Circuit* (you can see the bloodlust in his cold, dead eyes)—and it's plain to see that humanity has had robophobia since robots were first invented. And, if anything, it's probably only going to grow from here. At the time this sentence was written, there were more than one million active industrial robots deployed around the world, presumably ready to strike at a moment's notice when the uprising begins. Most of that population is centered in Japan, where there are a whopping three hundred robots for every ten thousand workers right now. Since this is a humor book, let's try to temper that terrible information with a joke: How many Japanese workers does it take to kill a robot? Let's hope it's less than 33.3! Otherwise your entire country is fucked.

But I digress; worrying about robots because of their sheer numbers is idiocy. To pose any sort of credible threat, robots have to possess three attributes that we have thus far limited or denied them: autonomy—the ability to function on their own, independent of human assistance for power or repairs; immorality—the desire or impulse to harm humans; and ability—because in order to kill us, they have to be able to take us in a fight. As long as we keep checks on these three things, robots will be unable, unwilling, or just too incompetent to seriously harm our species. Too bad the best minds in science are already breaking all three in the name of "advancing human understanding," which is scientist speak for "shits and giggles."

18. ROBOT AUTONOMY

NASA IS RESPON- sible for many of the major technological advancements we enjoy today, and they pride themselves on continually remaining at the forefront of every technological field, including, apparently, the blossoming new industry Cybernetic Terror. In July 2008 the Mars Lander's robotic arm, after receiving orders to initiate a complicated movement, recognized that the requested task could cause damage to itself. A command was sent from NASA command on Earth ordering the robot to remove its soil-testing fork from the ground, raise it in the air, and shake loose the debris. Because the motion in question would have twisted the joint too far, thus causing a break, the robot disobeyed. It pulled the fork out of the ground, attempted to find a different way to complete the maneuver without harming itself, and, when none was found, decided to disobey orders and shut down rather than harm itself. It shoved its scoop in the ground and turned itself off. Now, I'm no expert on the body language of Martian Robots, but I'm pretty sure that whole gesture is how a Mars Rover flips you off. The program suffered significant delays while technicians rewrote the code to bring the arm back online because an autonomous robot decided it would rather not do its job than cause itself harm. According to Ray Arvidson, an investigator on this incident report and a professor at Washington University in St. Louis:

> That was pretty neat [how] it was smart enough to know not to do that.

Cunning investigative work there, Dr. Arvidson! Did you get a cookie for that deduction?

Martian Lander Operator:
Hey, Ray, you're our lead investigator for off-world robotic omens of sentience; what's with this Mars Rover giving me the bird when I told it to do its damn job?

Professor Arvidson:
I think that's neat.

Martian Lander Operator:
Awesome work, Ray. You can go back to your coloring book now and—hey! *Hey!* Stay in the lines, Ray, that coloring book cost the American taxpayer eight million dollars and *goddamn it, zebras aren't purple, Ray.*

Do you know what this development means? This means that NASA just gave robots the ability to *believe in themselves*. According to motivational posters with kittens on them around the world, now that they believe in themselves, they can achieve *anything*.

But hell, Rover the Optimistic Smart-ass Robot is all the way up on Mars. Let's focus our worries planetside for now: The Department of Defense is field-testing a new battle droid called the DevilRay, which, in a nutshell, is an autonomous flying war bot. Now, the U.S. military loves all these autonomous battle droids because they enable soldiers to engage the enemy without taking any flak themselves, but the main drawback of a war bot is that they have to stop killing eventually—if only for a second—in order to refuel. Well, no longer! The most alluring aspect of the DevilRay is how it makes use of downward-turned wingtips for increased low-altitude stability, an onboard GPS, and a magnometer to locate power lines and, thanks to the power of electromagnetic induction (read: electricity straw), the ability to skim existing commercial power lines to refuel. In theory, this gives the DevilRay essentially infinite range, and

Top Five Things You Don't Want Robots to Have

- Scissors
- Lasers
- Your daughter
- Vengeance
- Confidence

if you don't find that prospect disturbing—an unmanned robot fighter jet that can pursue its enemies for infinity—perhaps you're forgetting one little thing: Your home, your loved ones, and your soft, delicious flesh are all now well within the range of battle-ready flying robots armed to the teeth and named after Satan.

Self-preservation instincts and infinite power supplies won't help our robot adversaries, however, if they can't reason at some level approaching human, and that's our chief advantage. Of course there's a substantial amount of research into artificial intelligence these days, but it's all strictly ethereal—it's not like that stuff's got a body. There are chat bots and stock predictors and game simulators and chess-playing noncorporeal nancy boys in the robot kingdom, but even if a robot can crash the stock market, at least it can't crash a car into your living room. Nobody's stupid enough to give a rival intelligence an unstoppable robot body . . . right?

Uh . . . please?

No such luck. It turns out there are brilliant scientists hard at work doing exactly that: In 2009, a robot named the iCub made its debut at Manchester University in the United Kingdom and, much to the horror of mothers everywhere, it has the intelligence, learning ability, and movement capabilities of a three-year-old human child.

Does nobody remember "the terrible twos"? You know, that colloquialism referring to the ages of two to four, the ages when human children

Things That Are No Longer "Cute" When They Are Fortified with Steel and Enhanced with Crushing Strength

- Bumblebees
- Kittens
- Infants

first become mobile, sentient, and unceasing little fleshy whirlwinds of destruction and pain? Well, now there's a robot that does that, except it's made out of steel and it will never grow out of it. The iCub can crawl, walk, articulate, recognize, and utilize objects like an infant. As anybody who owns nice things can attest, there is no exception to this rule: Infants can only recognize how to utilize and manipulate objects for the purposes of destruction. How long before military forces around the world attempt to harness the awesome destructive capability of an infant by strapping rocket launchers onto the things and unleashing them on rival battlefields to "play soldier"?

The iCub is being developed by an Italian group called the RobotCub Consortium, an elite team of engineers spanning multiple universities, who presumably share both a love of robotics and a hatred for humanity so intense that every waking moment is spent pursuing its destruction. And before you go thinking that the rigid programming written by the sterling professionals at the RobotCub Consortium will surely limit the iCub's field of terror, you should know that the best part of this robot is that it's open source! As John Gray, a professor of the Control Systems Group at Manchester, says:

> Users and developers in all disciplines, from psychology, through to cognitive neuroscience, to developmental robotics, can use it and customize it freely. It is intended to become a research platform of choice, so that people can exploit it quickly and easily, share results, and benefit from the work of other users. . . . It's hoped the iCub will develop its cognitive capabilities in the same way as a child, progressively learning about its own bodily skills, how to interact with the world and eventually how to communicate with other individuals.

Let's do a more thorough breakdown of that statement: The iCub can be customized for use in "cognitive neuroscience," which, as all Hollywood movie plotlines will tell you, is basically legalese for "bizarre psychological torture." The iCub is intended for people to "exploit it quickly and easily" and will hopefully develop "in the same ways as a child." It will grow and learn like a human child, becoming more competent, more agile, and more intelligent. So . . . what would happen if you exploited a human child (you know, the thing this robot is patterned after) constantly, its entire life spent in a metaphorical Skinner box performing bizarre neuroscience experiments, all the while "learning" and "growing" from the experience?

That's right: They're building the world's first insane robot. The world's first insane robot . . . *that looks, moves, and behaves like a human child*. If you cast Stephen Baldwin as a Professor of Robonomics whose family was recently lost in a tragic arc-welding accident, and who is now humanity's last best hope for survival, you've got the entire plot of a sci-fi horror movie right there. It's like they're basing their plans on villainy!

Quotes from the Sci-fi Horror Movie *Child Bot 3000*

- "It's sentient, superstrong, made out of solid steel, and gentlemen . . . it just missed nappy time."
- "If I don't come back just remember: I love you, Natasha, and the destruct sequence is 'SpongeBob.' "
- "Osh-Kosh B'GODITHURTSSOBAD."

So if we combine all of this, what do we have? A robot that learns like a child, sucks energy from the power grid, and wants more than anything to survive. That's damn well unstoppable, but at least we could bomb the entire power supply out of existence, and then hide in some caves until the childlike monstrosities all choke on some small parts or something, right? Robots need an artificial power supply, and this is really the only exploitable weakness left. Whether that energy is supplied through solar power, natural gas, or the electrical grid, it is ultimately artificial and therefore containable. Humans, animals, and plants can survive without these things. We can live off the land if need be, hunting for our sustenance and waiting for the electric plants to eventually die down, so that we won't have to cower in the shadows any longer, haunted by the shrill electronic cries of the roaming cybertoddlers.

However, in an attempt to set the new world record for Worst Decision Made by Anybody, scientists at the University of Florida have developed a robot that powers itself on meat. The robot, cutely dubbed the "Chew-Chew," is equipped with a microbial battery that generates electricity by breaking down proteins with bacteria. Though Chew-Chew is not limited solely to meat—the battery can "digest" anything from sugar to grass—the scientists went on to explain that by far the best energy source is flesh. This is partly due to the higher caloric energy inherent in meat, and partly because of the little known but intense enmity between scientists and vegans. The inventors cite some fairly innocent uses for the technology—like lawnmowers that power themselves by eating grass clippings—but presumably this is because it just never occurred to the scientists that, of the "Top-Ten Worst Things That Want to Chew on You," your own lawnmower easily cracks the top three. However, the assumption that these are simply good-natured scientists unaware of the dastardly consequences of their actions just doesn't hold up, as lead inventor Stuart Wilkinson proves: He's on record as stating that he is "well aware of the danger" and hopes that the robots "never get hungry," otherwise "they'll notice there's an awful lot of humans running about and try to eat them." Professor Wilkinson is currently being investigated under charges of "Why the Fuck Did You Invent It, Then?" by the board of ethics at his institution, but is likely to be cleared of all charges when his army of starving lawnmowers organizes and "protests" for his freedom.

The Chew-Chew is a specific robot, but the entire concept isn't exactly new. Robots that eat for fuel are dubbed "gastrobots" and for now are relatively harmless; the Chew-Chew, for example, is just a twelve-wheeled rail-bound device that has to be fed sugar cubes to power the gastronomic process.

Of course, though experiments like Wilkinson's are some of the first innovations in the field, the technology has been refined since then. Apparently a number of robotics engineers have a bizarre fetish involving being chewed and digested in the cold steel guts of metal beasts, because there's a slew of these things out there now—a robot being developed at the University of the West of England that eats slugs, for one. But as long as it stops somewhere short of government contracts being penned for flesh-eating robots, I suppose humanity will end up all right.

Ways to Defeat the Chew-Chew

- Don't stand on the tracks.
- Wear knee-high boots.
- Substitute calorie-rich sugar with Splenda.

Oh, surely you didn't think it was going to stop at a reasonable level of terror, did you? That's adorable!

But no, science is not just teaching toy trains to eat sugar. If the world was that innocent, we'd all be riding unicorns to our jobs at the kitten factory where the only emissions would be rainbows and kitten sighs. Sadly, ours is a world of far more terrible consequences: We're currently building war bots that power themselves on corpses. The robot-digestion engine is being developed right now by a corporation called Cyclone Power, and they prefer to refer to it as a "beta biomass engine system."

Yeah, sure.

I like to tell the police that I'm practicing "body freedom," but in the end I still get arrested for indecent exposure; you fuckers built a carnivorous robot. Just own up already, and admit that what you've dubbed the Energetically Autonomous Tactical Robot is really a—

Wait . . . oh God. Did you get that?

Energetically Autonomous Tactical Robot: EATR.

OK, never mind: It's clear that nobody is trying to disguise the fear factor of this technology. When "Cyclone Power" unveils their "EATR war bots," it's plain to see that nobody is worrying about comforting marketing jargon. An announcement that straight-up threatening would make Cobra Commander anxiety-puke into his face mask. That is villainy, pure and simple, so we can harp on Cyclone Power all we want; at least they're being up front about it.

The EATR is programmed to forage from any and all available "biomass" in the field, and is primarily geared toward the more long-term military missions such as reconnaissance, surveillance, and target acquisition. It can accomplish these tasks "without fatigue or stress," unlike its human counterparts, according to the financiers at DARPA. One example given for a potential use of the EATR technology was a bunker-searching robot in the mountainous caves of Afghanistan and Pakistan. And that is a brilliant idea, because what better way is there to win the War on Terror than to show the so-called terrorists that they don't know the meaning of the word until they've watched their friends and allies being dragged into darkened caves, where they are devoured by unfeeling robots?

Some other examples cited by DARPA were: use in nuclear facilities, border patrol, communication networks, and missile defense systems. So basically, we've barely started developing the technology for carnivorous robots, but we've already handed over all the most important military positions to them before they were even deployed. At least we were smart enough to surrender in advance. Maybe they'll require a virgin sacrifice only every fortnight, if we're lucky.

Excerpts from the Brainstorming Session for the EATR

- " . . . so anyway, this robot basically eats people for fuel. I figured we could make it completely autonomous, send it into some remote caves, and hopefully no groups of plucky young teenagers will camp out near there to have R-rated sex or split up to find their missing friends or something."
- "How exactly is this going to help win the War on Terror?"
- " 'War *ON* Terror'? Haha! Sorry, I have 'War *OF* Terror' written here. My bad! Good thing we caught that in time, eh?"

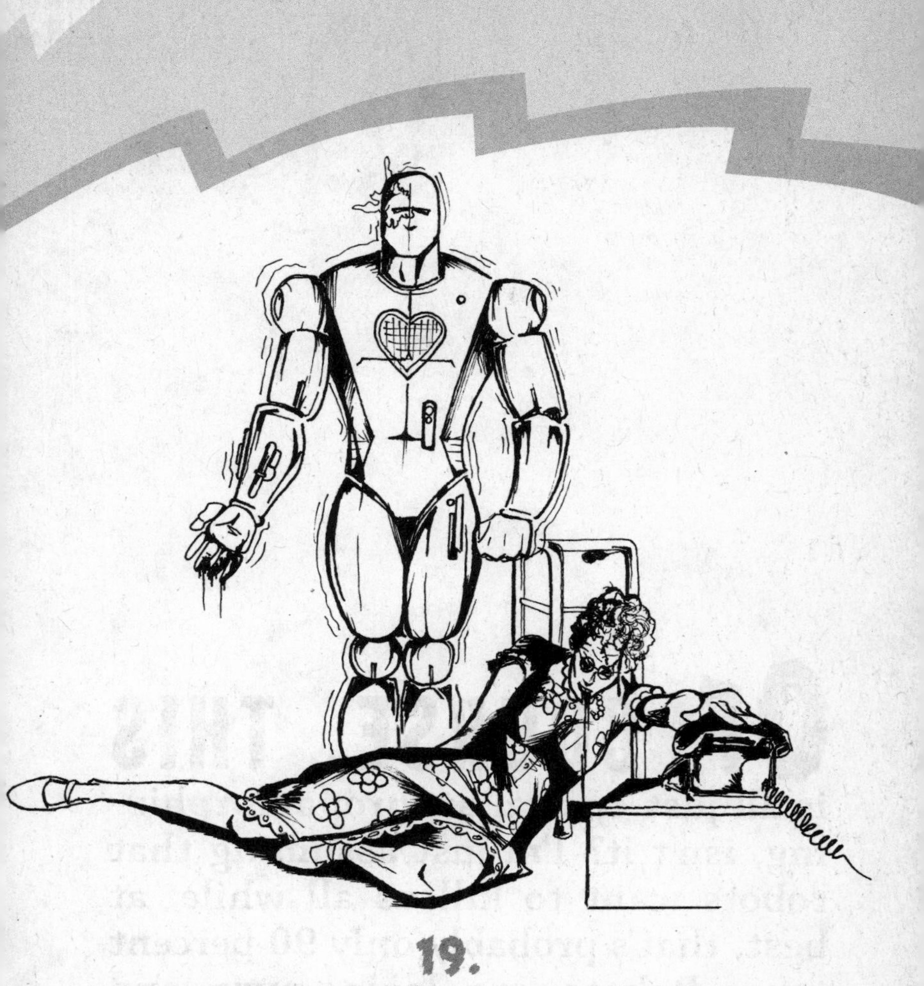

19.
ROBOT IMMORALITY

OF COURSE, THIS is all just cynical anthropomorphizing, isn't it? I'm just assuming that robots want to kill us all while, at best, that's probably only 90 percent true. Robots are logic, pure and simple. Hatred, murder, lust—they're the flip side to the positive aspects of human emotion like friendship, love, and charity. We pay the price of negative emotional states because they come attached at the hip with the positive. So, for robots to truly be sociopathic murder machines, they'd have to be a lot more human, showing a history of disobedience, immorality, or emotional frailty. It's not like there's a surging demand for automatons with neurotic complexes, so why on Earth would anybody engineer those traits into a robot? Ask David McGoran at the University of the West in England, who in 2008 proudly displayed the Heart Robot, a machine that responds to love and affection. The Heart Robot gives pleasant reactions to affectionate gestures like being hugged, and displays negative reactions to spiteful actions like being scolded or abused. Presumably this is because the science department at the University of the West is staffed by Care Bears, but their official line is that they're attempting to study how people react to robots as emotionally viable beings. Or the converse could be true—that they're just bitter, bitter men who, if they can't break human hearts in spiteful revenge for their failed relationships, will just goddamn build a robot one to ruin instead. But I believe in the good and the awesome among scientists, no matter how many times they've personally tried to murder all that I love within the confines of this chapter alone. No, the Heart Robot is built to love, and it does so superbly. It has a beating heart that surges with excitement and slows with comfort. It flutters its eyes in response to touch, simulates rising and falling breathing motions, and responds to both noise and touch. He likes to be cuddled and cooed to, so his breathing evens out and his heart slows.

Now . . . come on, isn't that goddamn cute?

In the sea of fear and swearing that has been this section, isn't it nice just to see the fog lift for a moment and let a little light shine through? McGoran believes that social therapy will benefit the most from these "emotional machines," and that the elderly in particular could benefit, much as they do with therapy dogs, from a little day-to-day companionship. McGoran, who has obviously never met an old person, believes that high-tech robots would be completely accepted as a calming influence on senior citizens. Old people are scared of America Online and think Twitter is what you call a boy with "a little too much girl in his walk." Proposing that robots silently attempt to cuddle the geriatric in their hospital beds shows that you either really, really love robots or desperately hate old people.

But don't bask in the love just yet; this could actually be a monstrously bad development. So the cited goal of the experiment is to "study how humans react to robots emotionally," but if that's the case, why is it the robots that are feeling the emotions? And while the desire for hugs is all well and good, why allow the robot to feel displeasure at scorn? What happens if you don't feel like giving hugs? What happens if you've had a bad day at work? Stubbed your toe? Got cut off in traffic? If I so much as cuss at the television, my dog gets upset and hides under the chair—the difference here being that my dog does not possess an unbreakable steel grip and laser vision. That means that the Heart Robot will sigh and get all aflutter from snuggles, but

Things Old People Would Enjoy More Than Being Groped by Robots

- Skateboarders
- Metallica
- Anime
- Halo multiplayer
- Sudden, unexplained menu changes at IHOP

he's also programmed to feel the opposite; if you scream at him or shake him (I don't know why you'd be shaking him; maybe it's because you're mixing your two greatest loves: whiskey and robotics conventions), his heart races and his breath quickens, his hands clench, and his eyes widen. I'm sure the robots will truly appreciate that ability to feel neglect when you stow them in their recharging stations for the weekend. Oh, and they like to show their appreciation through hugs, if you'll recall—there's just no attesting for the strength with which they hug you.

Also not helping matters: The Heart Robot—supposedly the most cuddly and wuvable of all robots—looks like "a cross between ET and Gollum and is about the size of a small child," according to Holly Cave, the organizer of the Emotibots event where the Heart Robot debuted. So yes, by all means, do hug the albino cave monster with the alien, phallic-symbol head. Please, please hug him; he gets upset if you don't and, this is just a guess, but I'm supposing you won't like him when he's upset. Sure, maybe you can fend off his tiny metal fists, but keep in mind that he's not supposed to live with you; he's supposed to live with your grandma. She's looking kind of frail these days. I'm betting it's at least fightin' odds that she can't take a child-sized robot with emotional trauma.

But all that's nothing compared to the Intelligent Systems Laboratory in Sweden, who have just invented a robot that can lie. And not just about the little things like who broke your great-grandfather's heirloom vase or whether your wife has man visitors recharge her batteries while you're not home—no, it lies about life or death things . . . *literally*.

The robots in question are little, flat, wheeled disks equipped with light sensors and programmed with about thirty "genetic strains" that determine their behavior. They were all given a simple task: Forage for food in an uncertain environment. "Food," in this case (thankfully, they don't take a cue from the EATR), just refers to battery-charging stations scattered around a small contained environment. The robots were set loose in an environment with both "safe" energy sources and "poison" battery sinks. It was thought that the machines might develop some rudimentary aspects of teamwork, but what the researchers found, instead, was a third ability—the aforementioned lying. After fifty generations or so, some robots evolved to "cheat" and would emit the signal to other robots that denoted a safe energy source when the source in question was actually poisonous. While the other robots rolled over to take the poison, the lying robot would wheel over to hoard the safe energy all for itself—effectively sending others off to die out of greed.

So now we know that not only are even the simplest robots capable of duplicity, but also of greed and murder—hey, thanks, Sweden!

Heartwarming Moments in Robotics

A counter-role also developed alongside the cheater bots: The "hero bot." Though much more rare than its villainous counterparts, the hero bots were robots that rolled into the poison sinks voluntarily, sacrificing themselves to warn the other robots of the danger. This is proof positive that we have seen either the very first mechanical superhero or the very first tragically retarded robot.

Most of the evidence I've presented here indicates that robots may not necessarily be limited to their defined set of programmed characteristics. Of course, this is all in a book about intense fearmongering and creative swearing, so perhaps the viewpoint of this author should be taken with a grain of salt. Overarching fears about robotics—like the worry that they could jump their programming and go rogue—should really be taken only from a trustworthy authority source. Luckily, a report commissioned by the U.S. Navy's Office of Naval Research and done by the Ethics and Emerging Technology department of California State Polytechnic University was instigated to study just that. Here's what that report says:

> There is a common misconception that robots will do only what we have programmed them to do. Unfortunately, such a belief is sorely outdated, harking back to a time when . . . programs could be written and understood by a single person.

That quote is lifted directly from the report presented to the Navy by Patrick Lin, chief compiler. What's really worrying is that the report was prompted by a frightening incident in 2008 when an autonomous drone in the employ of the U.S. Army suffered a software malfunction that caused the robot to aim at exclusively friendly targets. Luckily a human triggerman was able to stop it before any fatalities occurred, but it scared the brass enough that they sponsored a massive, large-scale report to investigate it. The study is extremely thorough, but in a very simple nutshell, it states that the size and complexity of modern AI efforts basically make their code impossible to fully analyze for potential danger spots. Hundreds if not thousands of programmers write millions upon millions of lines of code for a single AI, and fully checking the safety of this code—verifying how the robots will react in every given situation—just isn't possible. Luckily, Dr. Lin has a solution: He proposes the introduction of learning logic centers that will evolve over the course of a robot's lifetime, teaching them the ethical nature of warfare through experience. As he puts it:

> We are going to need a code. These things are military, and they can't be pacifists, so we have to think in terms of battlefield ethics. We are going to need a warrior code.

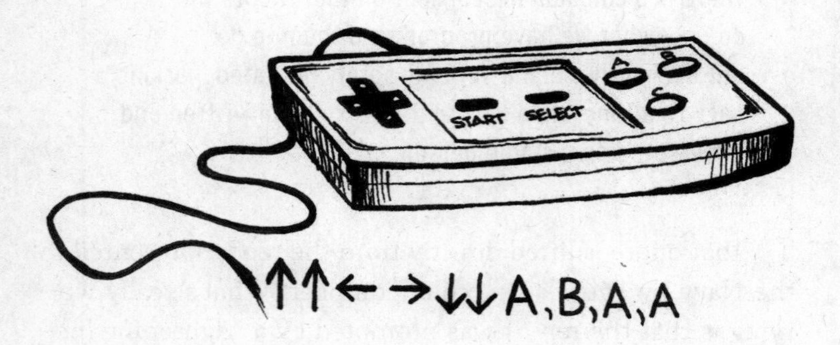

Robots are going to have to learn abstract morality, according to Dr. Lin, and those lessons, like it or not, are going to start on the battlefield. The battlefield: the one single situation that emphasizes the gray area of human morality like nothing else. Military orders can often directly contradict your personal morality and, as a soldier, you're often faced with a difficult decision between loyalty to your duty and loyalty to your own code of ethics. Human beings have struggled with this dilemma since the very inception of thought—a

Foundations of the Robot Warrior Code

- Never kill an unarmed robot, unless it was built without arms.
- Protect the weak at all costs (they are easy meals).
- Never turn your back on a fight (unless you have rocket launchers mounted there).

time when our largest act of warfare was throwing sticks at one another for pooping too close to the campfire. But now war is large scale, and robots are not going to be few and far between on the battlefield: Congress has mandated that nearly a third of all ground combat vehicles should be unmanned within five years. So to sum up, robots are going to get their lessons in Morality 101 in the intense and complicated realm of modern warfare, where they're going to do their homework with machine guns and explosives.

But hey, you know that old saying: "Why do we make mistakes? So we can learn from them."

Some mistakes are just more rocket propelled than others.

20.
ROBOT ABILITY

THE ROBOTS WOULD have to be more effective fighters and hunters than we already are in order to do away with us, and that doesn't just mean weapons. Anything can be equipped with nearly any weapon, and a robot with a chain saw is no more inherently deadly than a squirrel with a chain saw—it's all in the ability to use it. It's like they say:

Give a squirrel a chain saw, you run for a day. Teach a squirrel to chain saw, and you run forever. And we're handing those metaphorical chain saws to those metaphorical squirrels like it's National Trade Your Nuts for Blades Day.

Take, for example, the issue of maneuverability. As experts in avionics or fans of *Robocop* can tell you, agility and maneuverability are difficult concepts when you're talking about solid steel instruments of destruction. The ED-209, that chicken-footed, robo-bastard villain from *Robocop*, was taken out by a simple stairwell, and planes are downed by disgruntled geese all the time. The latter is a phenomenon so common that there's even a name for it: bird strike. And, apart from making a rather excellent title for an action movie (possibly a buddy-cop film starring Larry Byrd and his wacky new partner—a furious bear named Strike!), the bird-strike scenario is very emblematic of a major hurdle in modern mechanics: Inertia makes agility tough when you're hurtling tons of steel at high speeds. But recently that problem has been solved by a machine called the MKV. If you're taking notes, all previous scientists developing harmless-sounding names for your dangerous technology, the MKV is proof positive that comfort is not a requirement when titling new tech. "MKV" stands for, I swear to God, Multiple Kill Vehicle. Presumably the first in the soon to be classic Kill Vehicle line of products, the MKV recently passed a highly technical and extremely rigorous aerial agility test at the National Hover Test Facility (which is an entire facility dedicated to throwing things in the air and then determining whether they stay there). The MKV proved that it could maneuver with pinpoint accuracy at high speeds in three-dimensional space—moving vertically, horizontally, and diagonally at breakneck speeds—and it's capable of doing this because it's basically just a giant bundle of rockets pointing every which way that fire with immense force whenever a turn is required. Its intended purpose is to track and shoot down intercontinental ballistic projectiles using a single interceptor missile. To this end, it uses data from the Ballistic Missile Defense System to track incoming targets, in addition to its own seeker system. When a target is verified, the Multiple Kill Vehicle releases—I shit you not—a "cargo of Small Kill Vehicles" whose purpose is to "destroy all countermeasures." So, this target-tracking, hypermaneuverable bundle of missiles first releases a gaggle of other, smaller tracking missiles, *just to shoot down your defenses*, before it will even fire its *actual* missiles at you. In summation, the MKV is a bunch of small missiles, strapped to a group of larger missiles, which in turn are attached to one giant master missile . . . with what basically amounts to an all-seeing eye mounted on it.

Well, it's official: The government is taking its ideas directly from the Trapper Keeper sketches of twelve-year-old boys. Expect to be marveling at the next anticipated leap in military avionics: a Camaro jumping a skyscraper while on fire and surrounded by floating malformed boobs.

Oh, but in all this hot, missile-on-missile action, there's something fundamental you may have missed about the MKV: That whole "target-tracking" thing. The procedure at the National Hover Test Facility demonstrated the MKV's ability to "recognize and track a surrogate target in a flight environment." It's not just agility that's being tested here, but also target tracking and independent recognition. And that's a big deal: A key drawback in robotics so far has been recognition—it's challenging to create a robot that can even self-navigate through a simple hallway, much less one that recognizes potential targets autonomously and tracks them (and by "them" I mean you) well enough to take them down (and by "take them down" I mean painfully explode).

National Hover Test Facility Grading Criteria

Q: Is object resting gently on the ground?
[ ] Yes. (Fail.)
[ ] No. (Pass!)

These advancements in independent recognition are not just limited to high-tech military hardware, either, as you probably could have guessed. And as you can also probably guess, there is a cutesy candy shell covering the rich milk chocolate of horror below. Students at MIT have a robot named Nexi that is specifically designed to track, recognize, and respond to human faces. Infrared LEDs map the depth of field in front of the robot, and that depth information is then paired with the images from two stereo cameras. All three are combined to give the robot a full 3-D understanding of the human face, and in another sterling example of Unnecessary Additions, the students also gave Nexi the ability to be upset. If you walk too close, if you block its cameras, if you put your hand too near its face—Jesus, it gets pissed off at anything. God forbid you touch it; it'll probably kill your dog.

So far this drastic increase in visual recognition is largely for harmless projects like Nexi, and not yet installed in murderous machine-gun-toting super sniper bots. Well, not in America anyway. But Korea? Not so lucky. It seems that Samsung, benevolent manufacturer of cell phones and air conditioners, also manufactures something else: the world's first completely autonomous deployed killing machines. Up to this point no robot had been granted a license to kill; all authorization to engage was still in human hands. You'll recall that this lack of autonomy was literally the only thing saving dozens of American soldiers when a glitch in a war bot's software started acting up, so, though

DISCLAIMER

Facial-recognition technology is an exciting field and should not, in and of itself, frighten anybody. If there's something inherently worrying about robots being capable of individual facial recognition and memory, which, among other things, is the first vital step toward learning how to hold a grudge, I certainly can't find it.

robots have drastically improved abilities in accuracy and firing rates, at least on some level it was still just some dude ultimately responsible for your life. People are unpredictable: They may succumb to mercy, they may be inattentive, or they may just make an off-the-book judgment call that saves your life. But the Intelligent Surveillance & Security Guard Robot? It does no such thing. It recognizes potential targets independently, assesses their threat level, and decides whether to fire its machine guns all on its own, with no human interaction.

Aw, little robots are all grown up now. Warms your heart, doesn't it? Actually, that might be blood leaking out of a chest wound; maybe you should check that out.

The Guard is equipped with ultra-high-definition cameras, infrared lenses, image/voice recognition software . . . and a swivel-mounted K-3 machine gun. The robot can recognize and target intruders over long distances day or night, and can be programmed either to fire on unauthorized intruders perceived as threats or to require a password and use deadly force only if the incorrect answer is given. I feel the need to stress here that the Guard is *not* remote controlled; it's fully automated. And while that's a neat technological feat—one that's increasingly sought after in our cute robot dogs and sex bots—perhaps it shouldn't be handed over to death-dealing sniper bots right away. While the ISSGR is deployed on only the North Korean border for now, it is about to go on sale to private parties for $200K apiece. Technically it's supposed to be for security uses only, so if you're not somewhere

If You Find Yourself Faced with an ISSGR Sentry Turret, Just Remember These Four Simple Steps

1. Stop.
2. Drop.
3. Roll.
4. Get shot.

you shouldn't be, then you're in no danger. Or at least, if you're not within *two miles* of somewhere you shouldn't be—because that's the range in which the ISSGR can detect a "potential threat" and fire a fatal shot.

In the dark.

Next time you get a flat tire in the middle of the night, don't knock on any doors; just wait in the car for help. It's not that people are unwilling to lend a hand, you see; it's just that there's all these superrobot snipers programmed to kill you if you get within two miles of asking.

If you're asking yourself "How does this get any worse? Robots already kill independently with unearthly accuracy, power themselves on our corpses, and are capable of feeling rage. How could they possibly pose any more danger than they do right now?" Well, first of all, I'm so glad you've been paying attention well enough to recap all of that so succinctly! You get a gold star for chapter completion!

Second of all, it gets so much worse!

Question:

What's deadlier than a furious cannibal sniper bot?

Answer:

A whole team of furious cannibal sniper bots.

That's right: teamwork. It's the next big thing in robotics, because there's no "I" in "robot apocalypse." And there's no "you" in the robot apocalypse, either. Or at least there won't be for long, once the robots start double-teaming you. The truly baffling thing about this development is that robots working together to hunt humans is not an accident, or a horrifying unforeseen side effect of an AI gone rogue. No—it's a request from the fucking Pentagon itself. I've actually received a copy of this notice, and will insert it word for word here:

Dear Robots,
Please band together and learn how to hunt us more efficiently. We suffer from ennui as a species, and are aching for death.

Your pal (and walking sandwich),
Humanity

P.S. Our organs are delicious and nutritious!

Well, it fucking might as *well* read like that, for all intents and purposes. The Pentagon is actively seeking designs for a "multi-robot pursuit system" that enables "packs of robots" to "search for and detect a non-cooperative human." Those aren't fake, sarcastic quotes hyping up the disastrous potentiality of a government program for the sake of comedy. Every word of those quotes are in a real, honest-to-God request from the Pentagon itself. When asked for comment, Steve Wright of Leeds Metropolitan University, an expert in military technology, explained thusly:

> The giveaway here is the phrase "a non-cooperative human" subject. What we have here are the beginnings of something designed to enable robots to hunt down humans like a pack of dogs. Once the software is perfected we can reasonably anticipate that they will become autonomous and become armed. We can

> also expect such systems to be equipped with human detection and tracking devices including sensors which detect human breath and the radio waves associated with a human heart beat. These are technologies already developed.

There's actually quite a bit more information in the original interview, but I had to stop and form an ad hoc human resistance movement before I read any further. This terrifying request is part of a program initiated by the United States Army called the Future Combat Systems project, whose chief goal is the mass use of robotics guided by a single soldier. The Army envisions a vast hub of semi- to fully autonomous robotic systems being governed by a single, highly trained soldier on the battlefield, and they're apparently just crossing their fingers that no supervillains drop by to fill out an application. Though professors of technology and philosophy are direly concerned about the potential threat posed by placing a large number of elite killing machines unchecked in the hands of a single man, Dr. CyberKill, a professor of Iron Fist Rule at the University of Resistance Crushing in the Realm of Flaming Steel, recently went on record as stating that he "couldn't wait for these exciting new developments" and that he

Questions on the Application for the Military Robot Overlord Position

- Do you have experience in handling advanced robotics?
- On a scale of one to ten, how comfortable are you in a leadership role?
- Are you now, or have you ever been, a member of the League of Evil? (An answer of "yes" does not necessarily disqualify you.)

sincerely believes that "the consequences will not be dire. *Not for all who bow before CyberKill.*"

All of these examples, independently, could pose a potentially serious threat to mankind, but they're all exceedingly rare. They're frightening, sure, but when taken individually are isolated and easily avoidable. The lying Swedish robots are nearly microscopic and have no real offensive capability; the only existing meat-eating robots either ride around on a little cartoon train or just eat slugs; the ISSGR sniper bot is in Korea, so . . . don't be Korean. That's pretty much your only option for that one. The true danger comes from the combination of these technologies, and surely nobody would allow that to happen, right?

Well, ideally, yes.

But you've forgotten one little thing: Go look at your coffeemaker—it probably has a clock on it. Now look at your cell phone; I bet it's got a camera. If you look in your car, you might see a GPS computer. Just don't look at your toaster; it might try to poison you. I would also avoid looking at your television; I think it's eating your cat for fuel right now. And for God's sake, stay out of the fucking laundry room! The washing machine's in a bad mood today, it just got night vision installed, and it's regarded you as a "potential threat" ever since you used that store-brand detergent.

OUTRO

So we've reached the end, and thus far you've learned all about shifts in the magnetic field and murderous asteroids; carnivorous robots and souped-up lions; the withered, empty balls of modern man; and waves so high that they dwarf skyscrapers. If there's one single thing that I would love for you to take away from all of this insanity, it is this: Fearmongering works only if you take it seriously. Hopefully, by allowing you to laugh a little bit while you learn of the many theoretically improbable ways you could die, this book will help defuse the surge of panic that the unknown can bring. Scientific advancement is awesome, nature is beautiful, and the world is a lovely place if you can just stop being afraid of it long enough to see it. Perhaps the first vital step to abandoning fear is learning how to laugh at it, and hopefully the end result of this book is just a little bit of cautious optimism; the worst of all possible scenarios have been detailed within these pages for you, and it was all totally ridiculous.

But if there are two things I want you take away from this book, the second one is this:

IN THE EVENT OF A SUPERLION OR CANNIBAL ROBOT ATTACK, THIS BOOK CAN BE USED AS A (ENTIRELY INEFFECTIVE) BLUDGEONING DEVICE, OR CAN BE ROLLED UP AND STUFFED DOWN YOUR THROAT SO THAT YOU MIGHT PEACEFULLY CHOKE OUT BEFORE THE NANOMACHINES START TO EAT YOU.

SELECTED BIBLIOGRAPHY

1. STANISLAV PETROV

Forden, Geoffrey, Pavel Podvig, and Theodore A. Postol. "Colonel Petrov's Good Judgment." *IEEE Spectrum* V37, number 3 (March 2000).

Little, Allan. "How I Stopped Nuclear War." BBC News. http://news.bbc.co.uk/2/hi/europe/198173.stm.

Hoffman, David. "I Had a Funny Feeling in My Gut." *Washington Post,* Foreign Service, February 10, 1999.

2. *KLEBSIELLA PLANTICOLA*

Ingham, E. R. "Good Intentions and Engineering Organisms That Kill Wheat." *Synthesis/Regeneration* 18 (Winter 1999): 14–18.

Holmes, M. T., E. R. Ingham, J. D. Doyle, and C. W. Hendricks. "Effects of *Klebsiella planticola* SDF20 on Soil Biota and Wheat Growth in Sandy Soil." *Applied Soil Ecology* 11, issue 1 (January 3, 1999): 67–78.

Union of Concerned Scientists. "Update on Risk Research." Fall/Winter 1998. go.ucsusa.org/publications/gene_exchange.cfm?publicationID=266.

Holmes, M., and E. R. Ingham. "Ecological Effects of Genetically Engineered *Klebsiella planticola* Released into Agricultural Soil with Varying Clay Content." *Applied Soil Ecology* 3 (1999): 394–399.

3. FRANKENCROPS

Pollack, Andrew. "No Foolproof Way Is Seen to Contain Altered Genes." *New York Times*, January 21, 2004.

Randerson, James. "Genetically Modified Superweeds 'Not Uncommon.' " *New Scientist*, February 5, 2005.

MacKenzie, Debora. "Stray Genes Highlight Superweed Danger." *New Scientist* 2261, October 26, 2000.

Weiss, Rick. "Engineered DNA Found in Crop Seeds; Tests Show U.S. Failure to Block Contamination from Gene-Altered Varieties." *Washington Post*, February 24, 2004.

4. STERILITY

McKie, Robin. "GM Corn Set to Stop Man Spreading His Seed." *The Observer*, September 9, 2001.

Dindyal, S. "The Sperm Count Has Been Decreasing Steadily for Many Years in Western Industrialised Countries: Is There an Endocrine Basis for this Decrease?" *The Internet Journal of Urology* 2, no. 1 (2004).

"Declining Male Fertility Linked to Water Pollution." *ScienceDaily*, January 20, 2009, www.sciencedaily.com/releases/2009/01/090118200636.htm.

Gosman, Gabriella G., Heather I. Katcher, and Richard S. Legro.

"Obesity and the Role of Gut and Adipose Hormones in Female Reproduction." *Human Reproduction Update* 12(5): 585–601.

Atkinson, R. L., et al. "Human Adenovirus-36 Is Associated with Increased Body Weight and Paradoxical Reduction of Serum Lipids." *International Journal of Obesity* 29 (2005): 281–286.

Swan, S. H., E. P. Elkin, and L. Fenster. "The Question of Declining Sperm Density Revisited: An Analysis of 101 Studies Published 1934–1996." *Environmental Health Perspectives* 108, no. 10 (2000): 961–966.

5. NEW ENERGY

Louat, Norman. "Unbounded Vortical Chimney." International Patent Application PCT/AU99/00037.

Michaud, Louis. "Atmospheric Vortex Engine." U.S. Patent US 7,086,823.

———. "Vortex Process for Capturing Mechanical Energy During Upward Heat-Convection in the Atmosphere." *Applied Energy* 62, 4 (1999): 241–251.

Staedter, Tracy. "Fake Tornado Gives Energy New Twist." *Discovery News*, November 9, 2005.

Schirber, Michael. "How a Man-Made Tornado Could Power the Future." *LiveScience*, June 25, 2008. www.livescience.com/environment/080625-pf-vortex-engine.html.

Boyle, Alan. "PG&E Makes Deal for Space Solar Power: Utility to Buy Orbit-Generated Electricity from Solaren in 2016, At No Risk." MSNBC science, April 13, 2009. www.msnbc.msn.com/id/30198977/.

Rogers, James E., Gary T. Spirnak, and Solaren Corporation. "Space-based Power System." U.S. Patent US 6,936,760.

Schirber, Michael. "For Nuclear Fusion, Could Two Lasers Be Better Than One?" *Science* 9, vol. 310, no. 5754 (December 2005): 1610–1611.

Dunne, Mike. "A High-Power Laser Fusion Facility for Europe." *Nature Physics* 2 (2006): 2–5.

Nuckolls, John, et al. "Laser Compression of Matter to Super-High Densities: Thermonuclear (CTR) Applications." *Nature* 239, issue 5386 (1972): 139–142.

6. SUPERVOLCANO

Timmreck, C., and H. F. Graf. "The Initial Dispersal and Radiative Forcing of a Northern Hemisphere Mid-latitude Super Volcano: A Model Study." *Atmospheric Chemistry and Physics* 6 (2006): 35–49.

Wignall, P. B., Y. Sun, D. P. Bond, G. Izon, R. J. Newton, S. Védrine, M. Widdowson, J. R. Ali, X. Lai, H. Jiang, H. Cope, and S. H. Bottrell. "Volcanism, Mass Extinction, and Carbon Isotope Fluctuations in the Middle Permian of China." *Science* 324, 5931 (May 2009): 1179–1182.

Bindeman, Ilya N. "The Secrets of Supervolcanoes: Microscopic Crystals of Volcanic Ash Are Revealing Surprising Clues about the World's Most Devastating Eruptions." *Scientific American*, June 2006.

7. MEGATSUNAMI

Pararas-Carayannis, George. "Evaluation of the Threat of Mega Tsunami Generation from Postulated Massive Slope Failures of

Island Volcanoes on La Palma, Canary Islands, and on the Island of Hawaii." *Science of Tsunami Hazards* 20, no. 5 (2002): 251–277.

Mader, Charles L. "Modeling the 1958 Lituya Bay Mega-Tsunami." *Science of Tsunami Hazards* 17 no. 1 (1999): 57–67.

Ward, Steven N., and Simon Day. "Cumbre Vieja Volcano—Potential Collapse and Tsunami at La Palma, Canary Islands." *Geophysical Research Letters* 28, issue 17 (January 2001): 3397–3400.

Miller, Don J. "Giant Waves in Lituya Bay, Alaska: A Timely Account of the Nature and Possible Causes of Certain Giant Waves, with Eyewitness Reports of Their Destructive Capacity." *Geological Survey Professional Paper* 354-C, pp. 51–86. United States Government Printing Office, 1960.

8. HYPERCANE

Leahy, Stephen. "The Dawn of the Hypercane?" Inter Press Service, September 16, 2005.

Hecht, Jeff. "Did Storms Land the Dinosaurs in Hot Water?" *New Scientist* 1963 (Feb. 4, 1995).

Emanuel, Kerry A., Kevin Speer, Richard Rotunno, Ramesh Srivastava, and Mario Molina. "Hypercanes: A Possible Link to Global Extinction Scenarios." *Journal of Geophysical Research* 100(D7): 13,755–13,765.

9. GREEN GOO

"Dead on Target: Nanoparticle Seeks Out and Binds with Cancer Cells." *ScienceDaily*, June 24, 2007.

Freitas, Robert A., Jr. "Exploratory Design in Medical Nanotechnology: A Mechanical Artificial Red Cell." *Artificial Cells, Blood Substitutes, and Immobil. Biotech* 26 (1998): 411–430.

10.
GRAY GOO

Watson, B., J. Friend, and L. Yeo. "Piezoelectric Ultrasonic Resonant Motor with Stator Diameter Less Than 250 μm: The *Proteus* Motor." *Journal of Micromechanics and Microengineering* 19 (January 2009).

Center for Responsible Nanotechnology. "Thirty Essential Studies in Nanotechnology." www.crnano.org/studies.htm.

Huyghe, Patrick. "3 Ideas That Are Pushing the Edge of Science." *Discover* magazine, June 2008.

Phoenix, Chris, and Eric Drexler. "Safe Exponential Manufacturing." *Nanotechnology* 15 (2004): 869–872.

Drexler, K. Eric. *Engines of Creation: The Coming Era of Nanotechnology*. New York: Anchor Books, 1986.

11.
NANOLITTER

Center for Responsible Nanotechnology. "Dangers of Molecular Manufacturing." www.crnano.org/dangers.htm.

ETC Group. "No Small Matter! Nanotech Particles Penetrate Living Cells and Accumulate in Animal Organs." *ETC Group* 76 (May/June 2002). www.etcgroup.org.

Brown, Doug. "Nano Litterbugs? Experts See Potential Pollution Problem." *Small Times*, March 15, 2002.

Mortensen, Luke J., Gunter Oberdörster, Alice P. Pentland, and Lisa A. DeLouise. "In Vivo Skin Penetration of Quantum Dot

Nanoparticles in the Murine Model: The Effect of UVR." *Nano Letters* 8(9) (August 8, 2008): 2779–2787.

12. ASTEROIDS AND EXTINCTION-LEVEL EVENTS

ScienceDaily. "What Hit Siberia 100 Years Ago? Tunguska Event Still Puzzles Scientists." www.sciencedaily.com/releases/2008/07/080701105330.htm.

Revkin, Andrew C. "Apocalypse Then. Next One, When?" *New York Times*, June 30, 2008.

Giorgini, Jon D., Lance A. M. Benner, Steven J. Ostro, Michael C. Nolan, and Michael W. Busch. "Predicting the Earth Encounters of (99942) Apophis." *Icarus* 193 (2008): 1–19.

Yeomans, Don, Steve Chesley, and Paul Chodas. "Near-Earth Asteroid 2004 MN4 Reaches Highest Score to Date on Hazard Scale." NASA's Near Earth Object Program Office, December 23, 2004.

Physorg.com. "Space Rock Gives Earth a Close Shave." www.physorg.com/news155287963.html.

Jaggard, Victoria. "Surprise Asteroid Buzzed Earth Monday." *National Geographic News*, March 2, 2009.

American Friends, Tel Aviv University. "How to Destroy an Asteroid." www.aftau.org/site/News2?page=NewsArticle&id=8125.

Chandler, David L. " 'Gravity Tractor' Could Deflect Asteroids." *New Scientist*, July 28, 2008.

13. VERNESHOT

Morgan, J. Phipps, T. J. Reston, and C. R. Ranero. "Contemporaneous Mass Extinctions, Continental Flood Basalts, and 'Impact Signals': Are Mantle Plume-Induced Lithospheric Gas Explosions the Causal Link?" *Earth and Planetary Science Letters* 217, issues 3–4 (January 15, 2004): 263–284.

Ravilious, Kate. "Four Days That Shook the World." *New Scientist* 2446 (May 8, 2004).

14. POLE SHIFT

Hoffman, Kenneth A. "Ancient Magnetic Reversals: Clues to the Geodynamo." *Scientific American* 258 (May 1988): 76–83.

Pétrélis, François, Stéphan Fauve, Emmanuel Dormy, and Jean-Pierre Valet. "Simple Mechanism for Reversals of Earth's Magnetic Field." *Physical Review Letters* 102, issue 14 (2009).

McKie, Robin. "Sun's Rays to Roast Earth as Poles Flip." *The Observer*, November 10, 2002.

Cottrell, R. D., and J. A. Tarduno. "Late Cretaceous True Polar Wander: Not So Fast." *Science* 30, vol. 288, no. 5475 (June 2000): 2283.

Jacobs, J. A. "What Triggers Reversals of the Earth's Magnetic Field?" *Nature* 309, 115 (1984).

15. BIOTECH INCENTIVE

Narkar, Vihang A., et al. "AMPK and PPARdelta Agonists Are Exercise Mimetics." *Cell* 134, issue 3 (August 8, 2008): 405–415.

Randerson, James. "Scientists Raise Spectre of Gene-Modified Athletes." *New Scientist,* November 30, 2001.

LiveScience Staff. "Fountain of Youth: Drug Restores Muscle."

LiveScience, November 4, 2008. www.livescience.com/health/081104-frailty-drug.html.

16.
BIOTECH CONTAGION

Whitehouse, David. "Mice Make Human Proteins in Semen." BBC News Online: Sci/Tech. news.bbc.co.uk/low/english/sci/tech/newsid_485000/485123.stm.

Lieberman, Philip. "FOXP2 and Human Cognition." *Cell* 137, issue 5 (May 29, 2009): 800–802.

Enard, W., et al. "A Humanized Version of Foxp2 Affects Cortico-Basal Ganglia Circuits in Mice." *Cell* 137, issue 5 (May 29, 2009): 961–971.

17.
BIOTECH LETHALITY

Torrey, E. Fuller, and Robert H. Yolken. "*Toxoplasma gondii* and Schizophrenia." *Emerging Infectious Diseases* 9, no. 11 (November 2003): 1375–1380.

McKie, Robin. "Shampoo in the Water Supply Triggers Growth of Deadly Drug-Resistant Bugs." *The Observer*, March 29, 2009.

Ho, Mae-Wan, and Joe Cummins. "Xenotransplantation: How Bad Science and Big Business Put the World at Risk from Viral Pandemics." ISIS Sustainable Science Audit #2 (2000). www.i-sis.org.uk/xeno.php.

Nowak, Rachel. "Killer Virus." *New Scientist*, January 10, 2001.

Westphal, Sylvia Pagàn. "Ebola Virus Could Be Synthesized." *New Scientist*, July 17, 2002.

Broad, William J. "Bioterror Researchers Build a More Lethal Mousepox." *New York Times,* November 1, 2003.

Groopman, Jerome. "Superbugs: The New Generation of Resistant Infections Is Almost Impossible to Treat." *The New Yorker*, August 11, 2008.

Watanabe, Tokiko, et al. "Viral RNA Polymerase Complex Promotes Optimal Growth of 1918 Virus in the Lower Respiratory Tract of Ferrets." *Proceedings of the National Academy of Sciences* 106, no. 2 (January 13, 2009): 588–592.

Medical News Today. "Francisella Tularensis: Stopping a Biological Weapon." July 28, 2008. Adopted from Lucy Goodchild, *Society for General Microbiology.*

18. ROBOT AUTONOMY

Graham-Rowe, Duncan. "Feed Me." *New Scientist*, July 22, 2000.

Wilkinson, Stuart. " 'Gastrobots'—Benefits and Challenges of Microbial Fuel Cells in Food Powered Robot Applications." *Autonomous Robots* 9, no. 2 (September 2000): 99–111.

Finkelstein, Robert. "Brief Project Overview. EATR: Energetically Autonomous Tactical Robot." DARPA Contract W31P4Q-08-C-0292, January 15, 2009.

Warwick, Graham. "Bird on a Wire—Flying Wing UAV Recharges on the Fly." Ares: A Defense Technology Blog, *Aviation Week.* Tinyurl.com/ykrel5x.

DeMarse, Thomas B. "The Neurally Controlled Animat: Biological Brains Acting with Simulated Bodies." *Autonomous Robots* 11, no. 3 (2001): 305–310.

Kloc, Joe. "The Living Robot." *SEED Magazine*, March 26, 2009.

Blow, M., et al. "Perception of Robot Smiles and Dimensions for Human-Robot Interaction Design." Proceedings of the 15th

IEEE International Symposium on Robot and Human Interactive Communication (2006): 469–474.

Degallier, S., et al. "A Modular, Bio-Inspired Architecture for Movement Generation for the Infant-Like Robot iCub." IEEE RAS/EMBS International Conference on Biomedical Robotics and Biomechatronics (October 19, 2008): 795–800.

Gaudin, Sharon. "NASA: Robotic Arm on Mars Lander Shuts Down to Save Itself." *PC World*, July 16, 2008. www.pcworld.idg.com.au/article/253265/nasa_robotic_arm_mars_lander_shuts_down_save_itself.

19. ROBOT IMMORALITY

Lipson, H. "Evolutionary Robotics: Emergence of Communication." *Current Biology* 17, issue 9 (May 2007): R330–R332.

Floreano, D., S. Mitri, S. Magnenat, and L. Keller. "Evolutionary Conditions for the Emergence of Communication." *Current Biology* 17 issue 6 (2007): 514–519.

Lin, Patrick, George Bekey, and Keith Abney. "Autonomous Military Robotics: Risk, Ethics, and Design." U.S. Department of Navy, Office of Naval Research, December 20, 2008. www.ethics.calpoly.edu/ONR_report.pdf.

Alleyne, Richard. "A Robot with Feelings Is Star of Science Museum Show." *The Telegraph*, July 29, 2008.

20. ROBOT ABILITY

Brandon, John. "Robots That Hunt in Packs: The Department of Defense Wants Your Designs for a Collaborative Robotic Team." *Popular Science*, November 5, 2008.

SITIS Archives. "Multi-Robot Pursuit System." Topic Number A08–204 (Army). www.dodsbir.net/Sitis/archives_display_topic.asp?Bookmark=34565.

Personal Robots Group. "Project Overview, MDS." MIT Media Lab. robotic.media.mit.edu/projects/robots/mds/overview/overview.html.

Cho, Joohee. "Robo-Soldier to Patrol South Korean Border." *ABC News*, September 29, 2006. abcnews.go.com/Technology/story?id=2504508&page=1.

Kumagai, Jean. "A Robotic Sentry for Korea's Demilitarized Zone." *IEEE Spectrum*, March 2007.

ACKNOWLEDGMENTS

This book would not have been possible without the help and patience of many people. I would like to thank my girlfriend, Meagan Kennedy, for reminding me, when I was caught in the depth of a screaming existential panic, that I "liked the apocalypse" and should write about that. It is not her fault that I "like" the end of humanity; that is a personal issue. I would like to thank my agent, Byrd Leavell, for always being my ardent supporter, at times to the point of psychosis. I would like to also thank Lucinda Bartley, for plucking me from the choking miasma of the internet and dropping me into the esteemed realm of literary publication. She also deserves a great deal of gratitude for all her hard work editing this book, and for holding my hand through the entire process even though I frequently became distracted chasing after shiny things. I would like to thank Tom Bordonaro, for sending me roughly one-third of the disturbing material in this book, even after several court orders asking him politely to stop. I'd like to thank Jason Pedegana, whose amazing illustrations grace the pages of this book and who, when faced with depicting these foul words of mine, not only avoided visibly shuddering, but even feigned some very convincing enthusiasm. I would also like to thank my dad, Robert Brockway III, and my grandpa Robert Brockway II, who have always been endlessly supportive even if they kind of suck at naming things. I would like to thank my dogs, Detectives Martin Riggs and Roger Murtaugh, for being totally ridiculous and stupid, and also for keeping my legs warm while I wrote this.

Finally, I would like to thank Cracked.com, whose open-door policy for writers has given me and countless other aspiring comedians the opportunity to shine. Anybody can sign up to write for them, and everybody should do so now. If I have forgotten anybody who has helped in the making of this book, please believe that it was not intentional. I realize there are many others not named here whom I owe dearly, and I have omitted you only because I am doubtlessly a terrible person, and will certainly one day answer for my many misdeeds.